Iterative Methods for Diffractive Optical Elements Computation

Iterative Methods for Diffractive Optical Elements Computation

VICTOR SOIFER

Samara State Aerospace University, Russia

VICTOR KOTLYAR and LEONID DOSKOLOVICH

Image Processing Systems Institute, Russian Academy of Sciences, Samara, Russia

UK Taylor & Francis Ltd, 1 Gunpowder Square, London EC4A 3DE
USA Taylor & Francis Inc., 1900 Frost Road, Suite 101, Bristol, PA 19007

British Library Cataloguing in Publication Data

A catalogue record for this book is available from the British Library
ISBN 0-7484-0634-4 (cased)

Library of Congress Cataloging Publication Data are available

Cover design by Jim Wilkie

Typeset in Times 10/12pt by Keyset Composition, Colchester

Printed in Great Britain by T.J. International Ltd, Padstow

Contents

Preface

This book deals with novel iterative methods and algorithms for calculating the transmission function of phase diffractive optical elements (DOEs). The development of DOEs is a topical problem of interest to those involved in computer optics.

The DOE is a phase-only element whose superficial microrelief has a height comparable with the light wavelength used. The DOE may be implemented in the form of a transparency or a reflecting mirror. DOEs can operate over a wide range of wavelengths from UV to IR radiation. The lens, the zone plate, the diffraction grating, the array illuminator, the kinoform, the axicon, and the phase spatial filter are examples of DOEs.

The technology of DOE manufacturing involves, as a rule, the use of image generators such as laser photoplotters or e-beam lithographers. The most critical stage of the technological process of DOE fabrication is generally the solving of an ill-defined inverse problem. Note that the robustness of the solutions arrived at is of fundamental importance here. Such problems can most efficiently be tackled through the use of iterative methods.

A distinctive feature of this book is that the methods covered realize a number of new approaches to the problem of synthesis of optical components:

- the design of DOEs is treated as an ill-defined inverse problem of the scalar theory of diffraction
- the formulation of a well-defined problem is related to the use of regularization techniques
- the development of parametric iterative algorithms allows one to control the rate of convergence

- the interpolation of the pixels array of the basis DOE uses the pixels array of an additional DOE
- the iterative calculation of the DOE is capable of forming Gauss–Hermite, Gauss–Laguerre and Bessel light modes
- the iterative–analytical calculation of a multifocus DOE for focusing into a set of lines – the calculation employs a nonlinear transform of the phase function of a DOE focusing into a line and is computed via the ray-tracing approach
- the solution to the problem of the initial approximation for iterative algorithms – the initial guess is chosen on the basis of the ray-tracing solution.

All the methods discussed in the book have been verified through the numerical simulation. The fast Fourier transform algorithm gives a computational basis for all the methods considered.

Some of the algorithms have been studied comparatively in terms of their suitability for solving the same problem. For a number of the iterative algorithms, a rigorous proof to their convergence is given.

Note that all the methods presented are aimed at calculating highly effective DOEs forming desired images with a 60–100% energy efficiency.

The introduction gives a brief explanation of major concepts that are employed and detailed in the subsequent chapters. Chapter 1 deals with parametric iterative methods used for designing kinoforms and capable of yielding, via the successive approximation, a Fresnel or a Fourier approximation of nonlinear integral equations of optics. It is shown how to vary the rate of convergence of the iterative procedure by fitting the value of the regularization parameter. Chapter 2 is devoted to the adaptation of iterative algorithms for designing DOEs that have the radially symmetric transmission function, using a Hankel transform instead of a two-dimensional (2D) Fourier transform. Chapter 3 discusses the peculiarities of iterative algorithms in computing DOEs forming reference wavefronts. It is demonstrated that the wavefronts can be formed using amplitude masks. Chapter 4 presents iterative techniques for calculating DOEs forming Gaussian and Bessel modes. Light modes can propagate either all in one direction, thus forming a light beam with the required mode composition, or each in its own direction. Chapter 5 deals with techniques for iteratively calculating multi-order phase diffraction gratings with multilevel and binary phase. Chapter 6 presents analytic–iterative methods for calculating multifocus DOEs for focusing into a set of lines. The methods utilize the nonlinear transformation of the phase function of a basis DOE focusing into a line. Chapter 7 covers the problem of choosing the initial guess for iterative methods for designing DOEs and includes numerical results of a comparison between different methods in solving some special tasks: focusing into a transverse segment and axial segment, into a square, and into a ring. Appendices F and G treat the algorithms for synthesis of DOEs with regard for the electromagnetic radiation polarization.

In summary, the authors hope that this book will be of much help to optical engineers and systems designers. It can also be used as a textbook for senior students and postgraduates specializing in optics, digital holography, laser systems design, image processing, and phase retrieval, and is recommended for use as a manual for retraining courses, under the title 'Non-conventional optics for lasers'.

The authors wish to acknowledge that various studies pertaining to the contents of this book have been conducted at the Image Processing Systems Institute of the Russian Academy of Sciences, Samara, and Samara State Aerospace University.

We thank Drs S. N. Khonina, I. V. Nikolsky, P. G. Seraphimovich and S. V. Philippov, who undertook the computer-aided studies of DOEs, and Margarita I. Kotlyar for her help in translating the book into English. We also thank Jakov E. Takhtarov and Oleg K. Zalyalov for their time and effort in preparing a disk version of the book.

Introduction: Generation of Wavefields using Diffractive Optical Elements (DOEs)

I.1 Phase DOEs

A brief explanation is given here of the major concepts that are employed and detailed in the subsequent chapters.

Phase DOEs are the diffractive elements designed for implementing the required transformations of the light field without the loss of light energy, which means that the action of the phase DOE is limited to the modulation of the phase of incident light. Throughout this book, light is considered in terms of the scalar theory of diffraction, is perceived as being monochromatic and absolutely coherent, and is described by the complex function of one or two spatial variables. The module of this complex function is said to be the amplitude of the light field, and the argument is called the phase.

The propagation of light in free space can be described using the Kirchhoff diffraction approximation, and a mathematical aspect of light propagation through optical elements (lenses and DOEs) reduces to the multiplication of the complex function of incident light by the complex transmittance of an optical element.

Phase DOEs represent either a flat surface with microrelief (see Fig. I.1a) or a plane-parallel plate transparent for the light used and having a surface relief (see Fig. I.1b).

The term 'calculation of a DOE' is used to mean 'calculation of the phase function of a DOE'. Strictly speaking, the methods for iteratively calculating the phase function are the main subject of discussion in this book. The result of calculation is the function of two variables $\varphi(u, v)$ that have the meaning of DOE phase. An analytical formula for $\varphi(u, v)$ determines an optical surface that would theoretically be fabricated by conventional optical

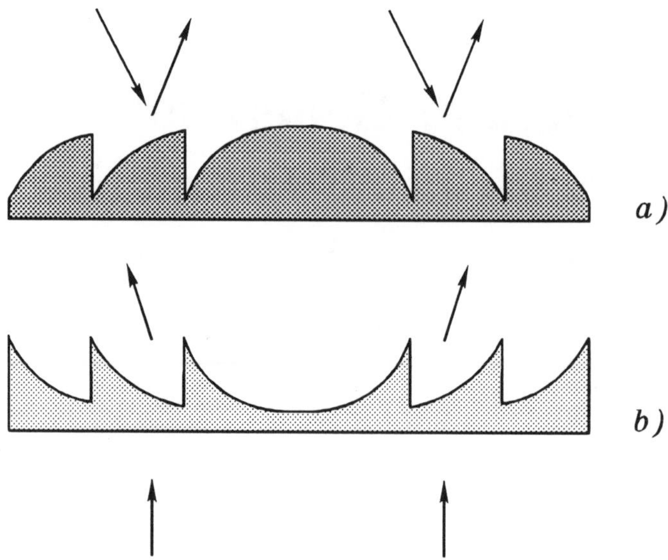

Figure I.1 Microreliefs of (a) a reflecting and (b) a transmitting DOE

techniques (cutting, milling, engraving, grinding, etc.). However, DOE technology similar to microlithographic technologies in terms of both concept and the equipment used is a matter of concern throughout this book. One of the main steps in this technology involves taking account of the periodicity of the phase $\varphi(u, v)$ and subsequently reducing its values to the interval $[0, 2\pi]$. In that case, the diffractive optical element only several wavelengths in microrelief height takes the place of the conventional optical elements of up to several thousand wavelengths in width. Figure I.2 shows the correspondence between the conventional optical element and the DOE.

It is seen that the relief on the DOE surface can have a regular zone structure. By way of illustration, the microrelief of a Rayleigh zone plate has a circular structure (see Fig. I.2b), whereas the microrelief of diffractive gratings is built of periodically located grooves. In the general case, the surface microrelief is irregular, as is the case for kinoforms and digital holograms [1,2], with zones to be described by curves (or fringes) which are more complicated than straight lines and circumferences. A distinction

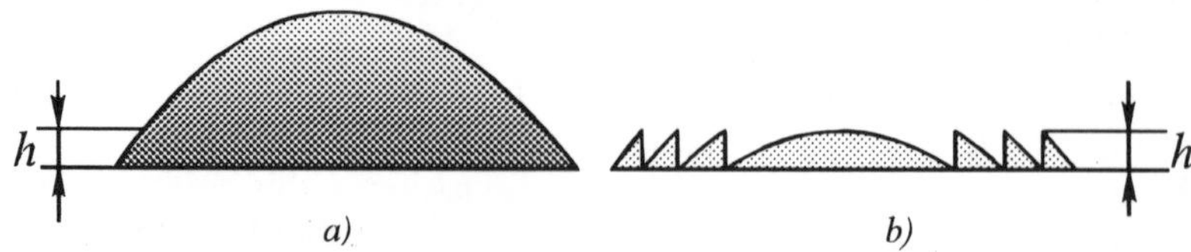

Figure I.2 An example of realization of the same phase function by means of (a) conventional and (b) diffractive optical elements

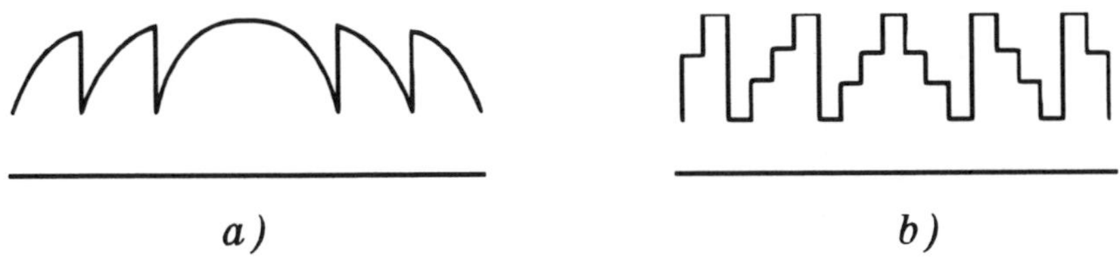

Figure I.3 (a) Continuous and (b) quantized microrelief of the DOE

between zone structures depends on how the DOE phase function was calculated. If the solution to the problem has been found in an analytical form, the DOE microrelief will be regular, whereas the use of an iterative procedure with a stochastic initial estimate in solving the problem will result in irregularity of the DOE microrelief and the structure of zones.

The DOE microrelief can be either continuous (see Fig. I.3a) or multilevel (quantized, see Fig. I.3b).

The depth of DOE microrelief is a multiple of the wavelength for which it is calculated. For example, for the reflecting DOE, the maximum relief height $h_{\max}$ (see Fig. I.3a) is estimated using the relationship

$$h_{\max} = \frac{\lambda m}{2\cos\theta}, \quad m = 1, 2, 3, \ldots \tag{I.1}$$

where λ is the light wavelength and θ is the angle of light incidence on the DOE surface. For the transmitting DOEs, instead of Eq. (I.1) the following equality should hold

$$h_{\max} = \frac{\lambda m}{n-1}, \quad m = 1, 2, 3, \ldots \tag{I.2}$$

where n is the refractive index of the DOE material.

The minimum modulation period of DOE microrelief (the minimum zone width) would be estimated in much the same way as the distance between the ring centres of neighbouring Fresnel zones [3], and is also specified by the angular size of the image formed by the DOE. Shown in Fig. I.4 is a phase DOE with radius R which, being illuminated by the light of wavelength λ, produces at the distance z a desired intensity distribution (image) characterized by the maximum linear size D in the transverse direction. Then, the minimum linear size $d_{\min}$ of modulation of the DOE microrelief can be found from

$$d_{\min} = \frac{2\lambda z}{2R + D} \tag{I.3}$$

The parameter $d_{\min}$ of Eq. (I.3) specifies the resolution of the technological equipment used in the fabrication of the DOE.

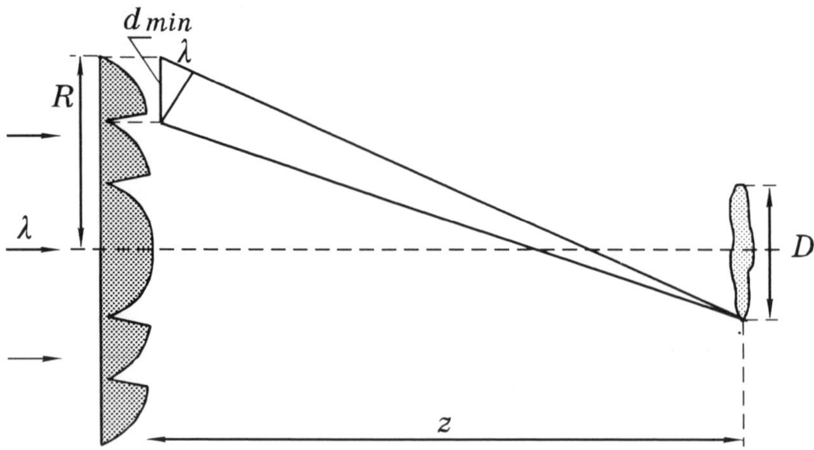

Figure I.4 Linear characteristics of the DOE and the image

The quality of the DOE calculation and fabrication is evaluated via the range of deviation of the produced image from the desired one and via the fraction of light energy going onto the generation of the image. Quality of the DOE is determined by the parameters

- *energy efficiency*, the ratio of the light energy having come into a pregiven area to the entire light energy striking the DOE
- *diffractive efficiency*, the ratio of the light energy having come into the pregiven area to the light energy that has passed through the DOE or reflected from it
- *image reconstruction error* (root-mean-square deviation), the square root of the sum of squares of the difference of the produced and desired intensity, with averaging summation performed over the entire domain of definition of the image.

In the field of image processing, where high accuracy of the formation of wavefronts or intensity distributions is required, the principal characteristic of the DOE quality is the image reconstruction error. In the area related to the material treatment using high-power laser radiation, the emphasis will be on the high energy density in the area of pregiven image. For problems like this, one needs to use phase DOEs providing a fairly high energy efficiency (close to 100%). Therefore, for calculating different-purpose DOEs different methods are used: for optical data processing, methods for coding digital holograms have proved their suitability; whereas for purposes of laser treatment, iterative methods for calculating kinoforms or ray-tracing methods for computing focusators have found wide acceptance.

Technological stages of the DOE fabrication are shown in Fig. I.5.

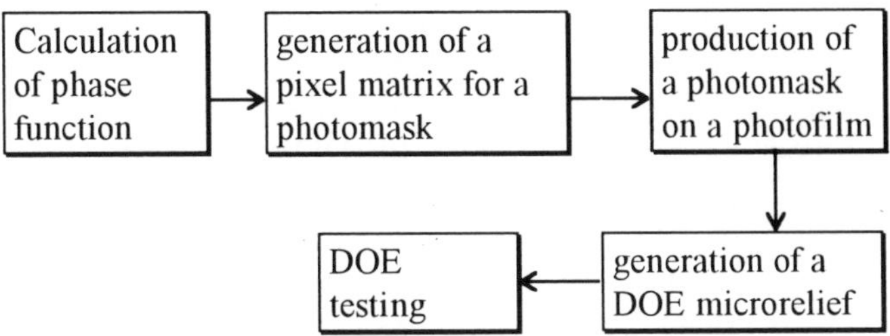

Figure I.5 Stages of DOE fabrication

In this book, we deal only with the first stage of this technological pattern: calculation of phase function. The essence of the subsequent stages can be briefly described as follows.

After the inverse problem has been solved and the DOE phase function has been derived on the basis of a set of desired characteristics of the image, one needs to transfer it onto a physical carrier as a microrelief. There are quite a lot of techniques for translating the phase function onto a DOE microrelief. Note, however, that a method of microlithography making it possible to obtain a stepwise microrelief [4] has found the widest use.

The starting-point for microlithography is a photomask that can be produced using a photo- or e-beam image generator and represents a binary amplitude mask consisting of transparent and opaque areas of photofilm. Next, the surface of a substrate is coated with photoresist and exposed to UV light through the photomask. Under the action of radiation, the photoresist changes its structure and, upon the succeeding chemical treatment, either illuminated or non-illuminated areas are removed, depending on the type of photoresist. As a result, windows arise in the photoresist layer through which the etching of the substrate material takes place. After etching and removal of the remaining photoresist, a microrelief corresponding to the phase pattern on the photomask will be found on the substrate. The microrelief depth depends on the time of etching. To produce a multistep microrelief, the above procedure should be repeated. Note that subsequent binary photomasks should be aligned precisely with the preceding ones. In such a manner one can fabricate transmitting phase DOEs. If a metal microlayer is deposited on the microrelief, a reflecting DOE will result.

It is also noteworthy that the process of DOE fabrication is, as a rule, iterative in nature since, if necessary, each stage is repeated or one returns to a preceding stage.

I.2 Methods for Calculating DOEs

Obtaining the phase transmission function of a DOE is always connected with solving an ill-posed inverse task. This task is inverse because it relates to problems of synthesis in which 'the source' should be found from arbitrarily specified 'data'. The problem of analysis is a straightforward problem in which an image synthesized by a pregiven 'source' is investigated.

Formally speaking, the calculation of the DOE phase is reduced to solving an integral equation. Incorrectness of the problem of synthesis lies in the fact that, first, there may not be a solution; second, if the solution exists it may not be unique; and third, the solution may be unstable. Instability of the solution to the inverse problem means that small deviations of the DOE phase from the calculated one may result in considerable deteriorations of the image formed.

As to non-uniqueness of the solution to the problem of DOE synthesis, as distinct from the field of image reconstruction and phase retrieval [5–7] in which non-uniqueness causes additional problems, in our case non-uniqueness proves to be useful and offers an opportunity to choose the best of several solutions.

In problems of image reconstruction, noise from the measurement data from which the object is reconstructed is responsible for incorrectness. In problems of DOE synthesis data noise is absent because image characteristics are specified uniquely. Note, however, that the desired image can be specified in such a manner that no solution to the inverse problem exists, which means that in the context of the formulated problem it appears to be impossible to calculate a phase DOE that would be able to generate the desired image accurately. That is why the procedure of regularization of the problem of DOE synthesis reduces to the replacement of the desired image by a similar one such that the solution exists and is stable.

The inverse problem of DOE synthesis is, generally speaking, nonlinear because it requires solving a nonlinear integral equation. The reason for nonlinearity is that one has to operate separately with the amplitude or with the phase of a complex function rather than with its real or imaginary part. In an early stage of development of digital holography, this nonlinear task was replaced by a linear one in which the amplitude–phase function of a hologram is derived as a result of conversion of linear integral Fresnel or Fourier transforms (see review in [8]). Next, the necessity of reducing the amplitude–phase function to the phase-only or the amplitude-only function required the elaboration of methods for coding: a method for coding binary amplitude Lohmann holograms [9,10], a Lee method [11], a method of parity sequences [12], etc. The Kirk–Jones method [13] of phase carrier represents a generalization of coding methods and allows a purely phase DOE to be calculated from the amplitude–phase function with a 30–40% efficiency.

The procedure of coding an amplitude–phase function via a phase-only function is carried out successively in the course of iteratively solving a nonlinear integral equation. For the first time, an iterative approach to solving the problem of DOE synthesis was suggested by Hirsh, Jordan and Lesem [2]. Gerchberg and Saxton [14] developed independently an analogous algorithm for reconstructing images. This provides the basic solution for the problem of synthesis of phase DOE, and is referred to below as an algorithm of error reduction or a Gerchberg–Saxton (GS) algorithm.

This algorithm has been elaborated in a series of subsequent works. In papers by Gallagher and Lin [15,16] the proof of the algorithm convergence is given. Chu and Fienup developed a parametric generalization of the GS algorithm and called it an input–output algorithm [17–19]. According to Fienup [20], the algorithm of error reduction is a special case of gradient methods. The presence of a parameter or a step in the iterative algorithm enables one to govern the convergence rate and to combat the stagnation effect typical of the GS algorithm. In papers by Broja, Bryngdahl, Weissbach and Wyrowski, an iterative error-diffusion algorithm for calculating DOEs is developed [21–23] and an iterative procedure of DOE phase quantization is given [24,25]. In papers by Kotlyar, Soifer and coworkers, an adaptive modernization of the GS algorithm is developed [26,27] and applied to computing radial DOEs [28,29] and iteratively to calculating phase formers of Bessel modes [30] and Gauss–Hermite modes [31,32].

An alternative approach to solving the problem of DOE synthesis is a ray-tracing method based on the analytical solution of the eikonal differential equation and on the construction of the ray path from the points on the DOE surface to the points of a desired image. This approach has been developed in papers by Danilov, Golub, Sisakyan, Soifer and coworkers [33–35]. Optical elements calculated via the above method have been fabricated for the IR and visible ranges of the spectrum and have successfully undergone tests [36–38]. Phase functions of DOEs calculated using the ray-tracing method are characterized by a regular zone structure because the eikonal differential equation can be solved analytically in an explicit form or using series.

As compared with coding methods [13] and a ray-tracing approach [34], iterative algorithms for solving integral equations [14] possess a number of advantages:

- a possibility of obtaining an approximate solution with desired accuracy
- a possibility of obtaining several solutions of the same problem by varying an initial estimate in the iterative process
- a combined solution to the inverse and straightforward problems in one algorithm, which means that a current estimate of DOE phase function derived in the inverse branch of the algorithm is checked and improved in the straightforward branch of the iterative algorithm
- universality and adaptability of iterative algorithms – universality is understood as independence from the kind of required image; adaptability is understood as suitability for computer-aided simulation and ease of programming
- a possibility of accounting for additional limitations imposed on the desired function of the DOE phase – introduction of a quantization or

sampling operator or other phase predistortions. Note that the process of development and application of novel iterative algorithms for calculating DOEs has not yet been finished, and the reader is encouraged to take part in it.

1.2.1 Peculiarities of Calculating Various Types of DOE

No algorithms are available to calculate all the possible types of DOE. For calculating each particular type of DOE one can select an optimal method. Phase DOEs can be conditionally divided into the following types:

1. DOEs generating half-tone images – digital holograms [8,11,39], kinoforms [2,16,19,25,26]
2. DOEs focusing laser radiation into a small spatial area – focusators [33–38]
3. DOEs producing binary images – diffraction gratings [40–42], matrix illuminators [43–46]
4. DOEs forming axial images – multifocus lenses [47,48], axicons [49,50]
5. DOEs generating wavefronts – correctors [51], compensators [52,53]
6. DOEs producing light modes – Bessel modes or nondiffracting beams [54,55], Gauss–Hermite and Gauss–Laguerre modes [56].

Let us take a brief look at the features of the above types of DOE.

1. For digital holograms intended for the generation of three-dimensional (3D) or 2D half-tone images one needs to calculate transmission functions as a matrix characterized by the large number of pixels $M \times M$ [57]. For example, the fabrication of a visually perceptible hologram 100×100 mm in size with a resolution of 1 μm calls for the calculation of a pixel matrix with $M = 100\,000$. In that case, the procedure of interpolation or extrapolation of a hologram datum array has proved successful [58]. In a simple situation, the extrapolation reduces to the replication of a calculated basic array of size $N \times N$, $N \ll M$. Note that in this case the hologram features the raster structure of phase, with the image consisting of a regular set of light spots (speckles).

Procedures of phase encoding [8], which can also perform an interpolation operation, because each pixel of the hologram is represented (encoded) as a rectangular sector of an amplitude [10] or phase [13] raster, have also found wide use in the calculation of digital holograms. It is noteworthy that iterative methods have also received wide acceptance in the computation of holograms and kinoforms [2,4,58].

2. Focusators are DOEs focusing laser light into a line with a pregiven energy distribution within this area [33–36]. Such optical elements have found application in laser technology for cutting, welding, branding, and hardening

materials [37]. The energy efficiency determines the quality of focusing and should not be less than 85–95%. The high DOE efficiency dictates the choice of the method of calculation. A ray-tracing method provides greater efficiency, but at some expense to the accuracy of formation of the focusing spot.

3. For computing DOEs aimed at generating the binary images of multiorder diffraction gratings or matrix illuminators employed in optical neuron networks, one uses methods for solving linear sets of algebraic equations [41], stochastic methods [59], and other methods of nonlinear optimization [40,43]. The requirement that the profile of such DOEs be binary determines the choice of a particular method. Another requirement is that the binary phase diffraction grating should form a number of orders more than 64×64, characterized by equal amplitudes with an 80% efficiency and with a spread of energies between orders of not more than 10%.

Recent years have seen a growth of interest in the synthesis of diffraction gratings with ultrashort period, which means a period less than or equal to the wavelength of the light used. The calculation of the microrelief of such surfaces is based on the exact solution of Maxwell equations [60,61]. The formulation of an inverse problem in this situation appears to be quite a challenging task because of the absence of an explicit analytical relation between the shape of the grooves of the diffraction grating and the intensity of light scattered. Note, however, that gratings with ultrashort period have found use as polarizing splitters of laser light [61] and for turning the light polarization vector [62].

4. DOEs forming axial images possess, as a rule, the radial structure of microrelief zones, which means that levels of equal phase are found on the concentric rings. The calculation of such DOEs requires the use of iterative methods involving algorithms of the fast Hankel transform [28]. In the course of calculation of radial DOEs able to form annular intensity distributions, an undesirable intensity peak is found in the ring centre. To avoid this, phase rotor masks are employed [63,64].

Note also that in a number of cases the problem of calculating radial DOEs can be reduced to that of computing one-dimensional (1D) DOEs [48]. Such elements have found use for the enhancement of focal depth of microlenses in laser-based reading in to/out of a disk [65], for pointing the laser beam at large distances [66], in bifocal microscopes [67], etc.

5. Phase formers of wavefronts are employed as correctors of aberration for conventional lenses [51], as compensators for checking aspherical mirrors [52] or as formers of reference wavefronts in the Twyman–Green interferometer [53]. The task of calculating such DOEs via iterative methods may be thought of as an inverse problem to that of computing kinoforms, because it requires the calculation of the DOE using the pregiven phase distribution in a certain plane, rather than the pregiven intensity distribution.

The quality of correctors is evaluated not so much via their energy efficiency as via the accuracy of formation of the pregiven wavefront.

The accuracy should be fairly high, up to tenths and hundredths of a wavelength.

6. Phase formers of light modes are employed for the purpose of ensuring light enters a set of fibres in a parallel manner or as spatial filters for analyzing the transverse-modal composition of laser light [56]. In calculating such DOEs, called modans, encoding methods such as a Kirk–Jones method [13] have found their use. The use of DOEs produced via such a method allows one to generate immediately and with high accuracy several Gauss–Hermite or Gauss–Laguerre modes propagating in different diffraction orders. However, one should note that the efficiency of such modans is only about 40%. The use of iterative algorithms for calculating DOEs capable of forming Gaussian modes allows a 90% efficiency [31,32] at a penalty of somewhat decreased accuracy in the required mode formation.

DOEs forming Bessel modes are employed for the generation of nondiffracting beams propagating in space without changing their shape [68] and for the investigation of properties of optical discharge in gases [69]. For the purpose of calculating the formers of Bessel modes, methods of coding [55] and iterative methods analogous to a GS algorithm [30] are utilized.

1

Parametric Methods of Computing DOEs

In this chapter we discuss iterative algorithms for solving a nonlinear integral Fresnel equation intended to compute phase optical elements forming an arbitrary pregiven intensity distribution of coherent monochromatic light in a certain plane perpendicular to the optical axis.

The light is said to be monochromatic and coherent if it can be described by a complex function satisfying the Helmholtz equation. Polarization effects are disregarded – the discussion is conducted in terms of the scalar theory of diffraction.

The algorithms under study are adaptive, because a new estimate of the sought function in each iteration step is chosen not only in accordance with the desired intensity function, but also depending on its previous estimate.

These algorithms are called 'parametric' because their convergence rate depends on the choice of particular values of some weight or regularization parameters.

In implementing the methods discussed, the fast 2D Fourier transform serves as a base computational algorithm.

1.1 Error Reduction Algorithm

Iterative algorithms [2,14] have found wide use in computing kinoforms [1] – phase optical elements dedicated to forming a desired light intensity distribution in some plane perpendicular to the optical axis.

In terms of the scalar theory of diffraction, the light complex amplitude in the optical element domain

$$W(u,v) = A(u,v)\,\mathrm{e}^{i\varphi(u,v)}$$

is related to the light complex amplitude

$$F(\xi, \eta) = B(\xi, \eta)\mathrm{e}^{i\psi(\xi,\eta)}$$

in the observation plane, in which the required intensity distribution $I_0(\xi, \eta)$ is formed, via the integral transform [70]

$$F(\xi, \eta) = \frac{ik}{2\pi z}\mathrm{e}^{ikz}\int_{-\infty}^{\infty}\int W(u, v)H(u - \xi, v - \eta, z)\,\mathrm{d}u\,\mathrm{d}v \qquad (1.1)$$

where

$$H(u, v, z) = \exp\left[\frac{ik}{2z}(u^2 + v^2)\right] \qquad (1.2)$$

is the function of the impulse response of free space in the Fresnel approximation, z is the distance between the DOE and the observation plane, and $k = 2\pi/\lambda$ is the light wavenumber for the wavelength λ. In Eq. (1.1), the complex amplitude $W(u,v)$ in the thin optical element approximation (transparency approximation), which disregards the ray refraction, is equal to the product of the complex amplitude $W_0(u, v)$ of illuminating light and the DOE's own transmission function $\tau(u, v)$

$$W(u, v) = W_0(u, v)\tau(u, v) \qquad (1.3)$$

Since in what follows we discuss only phase optical elements (unless otherwise specified), the DOE transmission is chosen as

$$\tau(u, v) = \mathrm{e}^{ig(u,v)} \qquad (1.4)$$

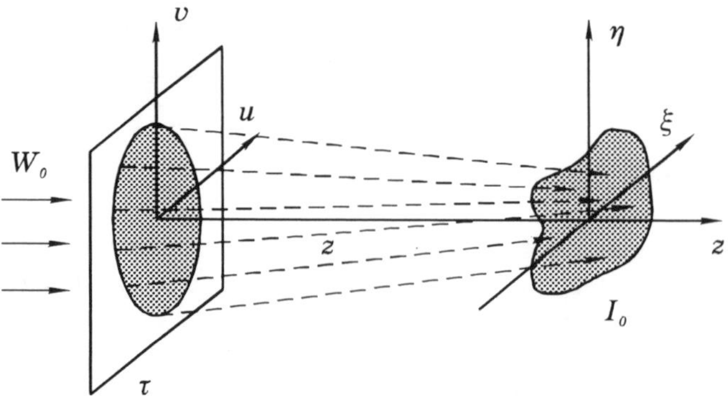

Figure 1.1 Schematic diagram of image generation using DOEs

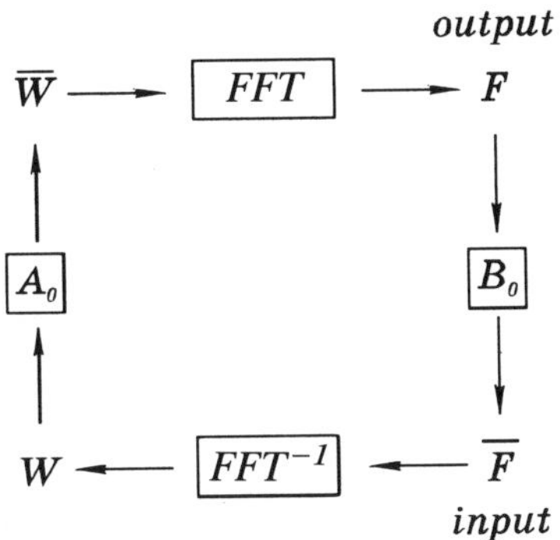

Figure 1.2 Flowchart of the error reduction algorithm

where $g(u, v)$ is the desired phase of the DOE. The notation is illustrated in Fig. 1.1. The problem of computing the DOE phase function $g(u, v)$ can be reduced to solving the nonlinear integral equation

$$I_0(\xi, \eta) = |F(\xi, \eta)|^2 = \left| \int_{-\infty}^{\infty}\!\!\int A_0(u, v)\,\mathrm{e}^{i\varphi(u,v)} H(u-\xi, v-\eta, z)\,\mathrm{d}u\,\mathrm{d}v \right|^2 \quad (1.5)$$

where $I_0(\xi, \eta)$ is the required intensity in the image plane, $A_0(u, v)$ is the amplitude of the illuminating beam, and $\varphi(u, v) = g(u, v) + g_0(u, v)$, where $g_0(u, v)$ is the illuminating beam phase. An iterative method of computing the phase $\varphi(u, v)$, and with it the DOE phase $g(u, v)$, consists of solving Eq. (1.5) via successive approximation. The Gerchberg–Saxton (GS) algorithm, or the error reduction algorithm [14], involves the following steps (see Fig. 1.2 for flowchart).

1. An initial phase guess $\varphi_0(u, v)$ is chosen.
2. The $A_0(u, v)\exp[i\varphi(u, v)]$ function is integrally transformed using Eq. (1.1).
3. The resulting complex amplitude $F(\xi, \eta)$ in the plane of image formation is replaced by $\overline{F}(\xi, \eta)$ by the rule

$$\overline{F}(\xi, \eta) = B_0(\xi, \eta) F(\xi, \eta) |F(\xi, \eta)|^{-1} \quad (1.6)$$

   where $B_0(\xi, \eta) = \sqrt{[I_0(\xi, \eta)]}$.
4. The inverse of the transform in Eq. (1.1) of the function $\overline{F}(\xi, \eta)$ is taken

$$W(u, v) = \frac{ik}{2\pi z}\mathrm{e}^{-ikz} \int_{-\infty}^{\infty}\!\!\int \overline{F}(\xi, \eta) H^*(\xi - u, \eta - v, z)\,\mathrm{d}\xi\,\mathrm{d}\eta \quad (1.7)$$

5. The resultant complex amplitude $W(u,v)$ in the DOE plane is replaced by $\bar{W}(u,v)$ according to

$$\bar{W}(u,v) = \begin{cases} A_0(u,v)W(u,v)|W(u,v)|^{-1}, & (u,v) \in Q \\ 0, & (u,v) \notin Q \end{cases} \tag{1.8}$$

where Q is the DOE aperture shape.

6. Passage to step 2.

This procedure is iterated until the errors δ_F and δ_W cease to vary significantly

$$\delta_F^2 = \frac{\int_{-\infty}^{\infty}\int [|F(\xi,\eta)| - B_0(\xi,\eta)]^2 \mathrm{d}\xi \mathrm{d}\eta}{\int_{-\infty}^{\infty}\int B_0^2(\xi,\eta)\mathrm{d}\xi \mathrm{d}\eta} \tag{1.9}$$

$$\delta_W^2 = \frac{\int_{-\infty}^{\infty}\int [|W(u,v)| - A_0(u,v)]^2 \mathrm{d}u \mathrm{d}v}{\int_{-\infty}^{\infty}\int A_0^2(u,v)\mathrm{d}u \mathrm{d}v} \tag{1.10}$$

The GS algorithm is also called the error reduction algorithm, since the errors of Eqs (1.9) and (1.10) have been demonstrated [15] to decrease with iterations. The GS algorithm has been shown to be a version of the gradient method [20] or of the gradient method of fastest descent [71]. This can be used to minimize the functional of the r.m.s. deviation of the reconstructed image amplitude from a pregiven one

$$\varepsilon_0 = \int_{-\infty}^{\infty}\int [|F(\xi,\eta)| - B_0(\xi,\eta)]^2 \mathrm{d}\xi \mathrm{d}\eta \tag{1.11}$$

Note, however, that the procedure of the GS algorithm convergence features a stagnation effect: in the course of several initial iterations the error δ_F (or δ_W) decreases rapidly, but subsequent iterations do not result in further significant reduction. To enhance the convergence rate of the procedure a variety of adaptive algorithms are employed, in which some parameters controlling the convergence rate are introduced.

A number of papers [18–20] develop an input–output method with reference to the problem of phase retrieval from the measurement of one intensity in the spatial spectrum plane.

The complex amplitude functions $W(u,v)$ (in the DOE plane) and $F(\xi,\eta)$ (in the Fourier plane in which the desired image is formed) are connected through a 2D Fourier transform

$$F(\xi,\eta) = \int_{-\infty}^{\infty}\int W(u,v)\exp\left[-i\frac{k}{f}(u\xi + v\eta)\right]\mathrm{d}u \mathrm{d}v \tag{1.12}$$

where f is the focal length of a lens forming the Fourier spectrum.

The change from the Fresnel transform, Eqs (1.1) and (1.2), to the Fourier transform, Eq. (1.12), means that in the last case one should employ a DOE–lens combination.

An iterative solution for the problem of light field phase reconstruction, viz. the $F(\xi, \eta)$ function argument, from the known amplitude, viz. the modulus $|F(\xi, \eta)|$, based on the input–output algorithm is given by

$$\bar{F}_{n+1}(\xi, \eta) = \bar{F}_n(\xi, \eta) + \beta\delta F_n(\xi, \eta) \tag{1.13}$$

where

$$\delta F_n(\xi, \eta) = \begin{cases} 0, & (\xi, \eta) \notin \gamma \\ -F_n(\xi, \eta), & (\xi, \eta) \in \gamma \end{cases} \tag{1.14}$$

$$F_n(\xi, \eta) = \mathfrak{F} D_A \mathfrak{F}^{-1}\{\bar{F}_n(\xi, \eta)\} \tag{1.15}$$

is the output function, $\mathfrak{F}$ and $\mathfrak{F}^{-1}$ are the Fourier transform and its inverse, D_A is a restricting operator in the DOE plane that can be exemplified by

$$D_A W_n = A_0 \frac{W_n}{|W_n|} \tag{1.16}$$

β is a parameter, and γ is a domain of violation of the restrictions imposed on the desired field function.

It is proposed that specifically for the problem of the DOE synthesis, the function increment δF_n in the input–output algorithm should be chosen in the form [19]

$$\delta F = [B_0 F|F|^{-1} - F] + [B_0 F|F|^{-1} - B_0 \bar{F}|\bar{F}|^{-1}] \tag{1.17}$$

where $B_0^2(\xi, \eta)$ is the desired intensity distribution in the Fourier plane, and $\bar{F}$ and F are the input and output functions, or the complex amplitude functions in the Fourier-image plane after and prior to the fulfilment of the restrictions.

Reference [73] reports a new version of the iterative input–output algorithm aimed at synthesizing DOEs, which assumes that the phase is independent of the module of the desired light amplitude $F(\xi, \eta)$. An optimal choice of the module and the phase in the $(n+1)$th iteration step is shown to take the form $(\bar{F}(\xi, \eta) = \bar{F})$

$$\begin{cases} |\bar{F}_{n+1}| = B_0 + \beta[|F_n| - B_0] \\ \arg \bar{F}_{n+1} = \arg F_n + \alpha[\arg F_n - \arg \bar{F}_n] \end{cases} \tag{1.18}$$

Based on numerical examples, it has also been shown in [73] that the best value of α and β is 2. Note that an iterative algorithm that employs the first equation in (1.18) is similar to the earlier algorithm described in [26,27].

1.2 Adaptive–Additive Algorithm

The input–output algorithm [18–20] was initially applied to reconstructing the phase of the light field from the measurements of one intensity in the spatial frequency plane. Such a problem is typical of stellar interferometry. The same algorithm was successfully employed for computing the DOE phase [19]. However, this algorithm lacks theoretical substantiation.

In the following we give a substantiation of a variant of input–output algorithms, which is adapted to calculating DOEs and is called an adaptive–additive (AA) algorithm. We have also proved the AA algorithm convergence (see Appendix A).

Let us consider the replacement for the GS algorithm in Eq. (1.6) in more detail. This replacement implies that the light amplitude $|F_n(\xi,\eta)|$ in the observation plane derived in the nth iteration step is replaced by a desired image amplitude $B_0(\xi,\eta)$. Note that although the Fourier (Fresnel) transform of a function with limited carrier is an analytical integer function of exponential type, the $B_0(\xi,\eta)$ function may be given arbitrarily and not represent an analytical function.

Therefore, it makes sense, instead of the replacement in Eq. (1.6), to try the replacement in which both functions (the desired one and the computed analytical one) participate as a linear combination with different weights [72]

$$|\bar{F}_n(\xi,\eta)| = \lambda B_0(\xi,\eta) + (1-\lambda)|F_n(\xi,\eta)| \tag{1.19}$$

In this case the replacement in Eq. (1.6) in the iterative GS algorithm takes the form

$$\bar{F}_n(\xi,\eta) = |\bar{F}_n(\xi,\eta)|F_n(\xi,\eta)|F_n(\xi,\eta)|^{-1} \tag{1.20}$$

The range of values for the λ parameter is found from the condition that the mean deviation ε_0 decided by the relation in Eq. (1.11) does not increase when one employs the replacement in Eq. (1.19). That is, we assume the fulfilment of the relation

$$\begin{aligned}\bar{\varepsilon}_0 &= \int_{-\infty}^{\infty}\!\!\int [|\bar{F}_n| - B_0]^2 \mathrm{d}\xi \mathrm{d}\eta = \int_{-\infty}^{\infty}\!\!\int [\lambda B_0 + (1-\lambda)|F_n| - B_0]^2 \mathrm{d}\xi \mathrm{d}\eta \\ &= (1-\lambda)^2 \int_{-\infty}^{\infty}\!\!\int [|F_n| - B_0]^2 \mathrm{d}\xi \mathrm{d}\eta \le \varepsilon_0 = \int_{-\infty}^{\infty}\!\!\int [|F_n| - B_0]^2 \mathrm{d}\xi \mathrm{d}\eta\end{aligned} \tag{1.21}$$

from which it follows that the weight parameter should be chosen from the condition $0 \le \lambda \le 2$. It is noteworthy that for $\lambda = 1$ the replacement in Eq. (1.19) changes to Eq. (1.6) for the GS algorithm, whereas $\lambda = 2$ gives a 'mirror image' replacement

$$|\bar{F}_n(\xi,\eta)| = |2B_0(\xi,\eta) - |F_n(\xi,\eta)|| \tag{1.22}$$

In the last case ($\lambda = 2$), the amplitude $|F_n(\xi,\eta)|$ in the observation plane computed in the nth iteration step is replaced by its 'mirror image' relative to the preset amplitude distribution $B_0(\xi,\eta)$.

Note that originally the 'mirror image' replacement in Eq. (1.22) was used in the other form [26,27]

$$\bar{I}_n(\xi, \eta) = \begin{cases} |2I_0(\xi, \eta) - I_n(\xi, \eta)|, & (\xi, \eta) \in \Omega \\ I_n(\xi, \eta), & (\xi, \eta) \notin \Omega \end{cases} \tag{1.23}$$

where I_n is the intensity distribution derived in the nth step, $I_n = |F_n(\xi, \eta)|^2$, I_0 is a pregiven intensity distribution, $I_0 = B_0^2$, and Ω is the domain of image definition. If $I_n \geq 2I_0$, $\bar{I}_n = 0$.

Let us show that the replacement in Eq. (1.19) in the AA algorithm minimizes the residual functional ε_1, which represents the r.m.s. deviation of the computed amplitude from the required one

$$\varepsilon_1 = \int_{-\infty}^{\infty}\int [A_0(u, v) - |W(u, v)|]^2 \mathrm{d}u\,\mathrm{d}v \tag{1.24}$$

where $A_0(u, v)$ is the light amplitude illuminating the DOE. The functional ε_1 variation relative to the function W in the nth step is given by

$$\delta\varepsilon_1 = 2\,\mathrm{Re}\left\{ \int_{-\infty}^{\infty}\int [W_n - A_0 W_n |W_n|^{-1}]\,\delta W_n^* \,\mathrm{d}u\,\mathrm{d}v \right\} \tag{1.25}$$

where Re{ } denotes the real part of the number.

In deducing Eq. (1.25), we made use of

$$\delta \int_{-\infty}^{\infty}\int A_0^2(u, v)\,\mathrm{d}u\,\mathrm{d}v = 0 \tag{1.26}$$

$$\delta \int_{-\infty}^{\infty}\int |W(u, v)|^2 \mathrm{d}u\,\mathrm{d}v \neq 0 \tag{1.27}$$

$$\delta(WW^*) = W\delta W^* + W^*\delta W = 2\,\mathrm{Re}\{W\delta W^*\} \tag{1.28}$$

Using the Parseval equation

$$\int_{-\infty}^{\infty}\int |W(u, v)|^2 \mathrm{d}u\,\mathrm{d}v = \int_{-\infty}^{\infty}\int |\bar{F}(\xi, \eta)|^2 \mathrm{d}\xi\,\mathrm{d}\eta \tag{1.29}$$

$$\int_{-\infty}^{\infty}\int |\bar{W}(u, v)|^2 \mathrm{d}u\,\mathrm{d}v = \int_{-\infty}^{\infty}\int |F(\xi, \eta)|^2 \mathrm{d}\xi\,\mathrm{d}\eta \tag{1.30}$$

where $\bar{W} = D_A W = AW|W|^{-1}$ is the light amplitude in the DOE plane after making the replacement of Eq. (1.8). One can verify that the following relation holds

$$\begin{aligned} &\mathrm{Re}\left\{ \int_{-\infty}^{\infty}\int [W_n - A_0 W_n |W_n|^{-1}]\,\delta W_n^* \,\mathrm{d}u\,\mathrm{d}v \right\} \\ &= \mathrm{Re}\left\{ \int_{-\infty}^{\infty}\int [\bar{F}_n - F_n]\,\delta \bar{F}_n^* \,\mathrm{d}\xi\,\mathrm{d}\eta \right\} \end{aligned} \tag{1.31}$$

where $\bar{F}_n = \mathfrak{F}\{W_n\}$

$$F_n = \mathfrak{F}D_A\mathfrak{F}^{-1}\{\bar{F}_n\} = \mathfrak{F}\{A_0 W_n |W_n|^{-1}\}$$

and $\mathfrak{F}$ denotes a Fourier transform.

The increment $\delta\varepsilon_1$ of the error functional ε_1 attains its maximum negative value in the nth step under the parametrically expressed condition

$$\delta\bar{F}_n = \bar{F}_{n+1} - \bar{F}_n = \lambda(F_n - \bar{F}_n), \qquad \lambda > 0 \tag{1.32}$$

In view of Eq. (1.32), instead of Eq. (1.25) we obtain

$$\delta\varepsilon_1 = -2\lambda \int_{-\infty}^{\infty}\!\!\int |\bar{F}_n - F_n|^2 \mathrm{d}\xi \mathrm{d}\eta < 0 \tag{1.33}$$

From Eq. (1.32) follows the parametric equation for iteratively deriving the $F(\xi, \eta)$ function

$$\bar{F}_{n+1} = (1-\lambda)\bar{F}_n + \lambda F_n = (1-\lambda)\bar{F}_n + \lambda\mathfrak{F}D_A\mathfrak{F}^{-1}\{\bar{F}_n\} \tag{1.34}$$

Equation (1.34) implies that in the $(n+1)$th step the input amplitude is a linear combination of the input and output amplitudes of the nth step. The algorithm of Eq. (1.34) is analogous to the input–output algorithm [20], and the preceding reasoning should be considered as its theoretical substantiation.

The algorithm of Eq. (1.34) converges, given that the λ parameter satisfies the inequality

$$0 < \lambda \leq 2 \tag{1.35}$$

Introduction of the notation

$$F_n = B_n \mathrm{e}^{i\varphi_n},\ \bar{F}_n = \bar{B}_n \mathrm{e}^{i\bar{\varphi}_n},\ W_n = A_n \mathrm{e}^{i\psi_n},\ \bar{W}_n = A_0 \mathrm{e}^{i\psi_n} \tag{1.36}$$

yields a sequence in which relations 1, 3, and 5 are evident, relations 2 and 4 are the Parseval equalities, and the inequality is that of triangle ($|a| - |b| \leq |a - b|$)

$$\begin{aligned}
\int_{-\infty}^{\infty}\!\!\int |A_0 - A_{n+1}|^2 \mathrm{d}u\mathrm{d}v &= \int_{-\infty}^{\infty}\!\!\int |A_0 \mathrm{e}^{i\psi_{n+1}} - A_{n+1}\mathrm{e}^{i\psi_{n+1}}|^2 \mathrm{d}u\mathrm{d}v \\
&\leq \int_{-\infty}^{\infty}\!\!\int |A_0 \mathrm{e}^{i\psi_n} - A_{n+1}\mathrm{e}^{i\psi_{n+1}}|^2 \mathrm{d}u\mathrm{d}v \\
&= \int_{-\infty}^{\infty}\!\!\int |B_n \mathrm{e}^{i\varphi_n} - [(1-\lambda)\bar{B}_n \mathrm{e}^{i\bar{\varphi}_n} + \lambda B_n \mathrm{e}^{i\varphi_n}]|^2 \mathrm{d}\xi\mathrm{d}\eta \\
&= (1-\lambda)^2 \int_{-\infty}^{\infty}\!\!\int |B_n \mathrm{e}^{i\varphi_n} - \bar{B}_n \mathrm{e}^{i\bar{\varphi}_n}|^2 \mathrm{d}\xi\mathrm{d}\eta \\
&= (1-\lambda)^2 \int_{-\infty}^{\infty}\!\!\int |A_0 \mathrm{e}^{i\psi_n} - A_n \mathrm{e}^{i\psi_n}|^2 \mathrm{d}u\mathrm{d}v = (1-\lambda)^2 \int_{-\infty}^{\infty}\!\!\int |A_0 - A_n|^2 \mathrm{d}u\mathrm{d}v
\end{aligned} \tag{1.37}$$

From Eq. (1.37) it follows that under the condition of Eq. (1.35) the algorithm converges on average to

$$\int_{-\infty}^{\infty}\int |A_0 - A_{n+1}|^2 \mathrm{d}u\,\mathrm{d}v \leq \int_{-\infty}^{\infty}\int |A_0 - A_n|^2 \mathrm{d}u\,\mathrm{d}v \qquad (1.38)$$

Note that the algorithm of Eq. (1.34) takes due account of the fact that the complex amplitude satisfies the restriction in the DOE plane, but not of the fact that the modules of the input and output functions of the complex amplitude in the Fourier plane, namely $\bar{F}(\xi, \eta)$ and $F(\xi, \eta)$, must tend to the pregiven function $B_0(\xi, \eta)$.

To employ this condition in the algorithm of Eq. (1.34) we assume that in the nth iteration step the input function obeys the restriction

$$\bar{F}_n(\xi, \eta) = D_B F(\xi, \eta) = B_0(\xi, \eta) F(\xi, \eta) |F(\xi, \eta)|^{-1} \qquad (1.39)$$

and in the $(n+1)$th iteration step it obeys Eq. (1.34). Therefore, a combination of Eqs (1.39) and (1.34) leads to the AA algorithm

$$\bar{F}_{n+1}(\xi, \eta) = (1-\lambda) B_0(\xi, \eta) \frac{F_n(\xi, \eta)}{|F_n(\xi, \eta)|} + \lambda F_n(\xi, \eta) \qquad (1.40)$$

Note that Eq. (1.40) differs from Eq. (1.19) by the permutation of the weight coefficients: $\lambda \to 1-\lambda$.

Convergence on average for the algorithms of Eq. (1.19) or (1.40) is proved in Appendix A. It is noteworthy that a similar technique was employed in [73] for deriving the algorithm of Eq. (1.40) from the algorithm of Eq. (1.34).

In [74] the algorithm (1.19) was obtained as a weight algorithm. Let us consider the residual functional ε_0, but with the weight function given by

$$\bar{\varepsilon}_0 = \int_{-\infty}^{\infty}\int S(\xi, \eta)[|F(\xi, \eta)| - B_0(\xi, \eta)]^2 \mathrm{d}\xi\,\mathrm{d}\eta \qquad (1.41)$$

where $S(\xi, \eta)$ is a real positive definite weight function that specifies, for example, the domain of definition for the image under synthesis

$$S(\xi, \eta) = \begin{cases} \lambda, & (\xi, \eta) \in \Omega \\ 0, & (\xi, \eta) \notin \Omega \end{cases} \qquad (1.42)$$

where λ is constant.

The variation of the functional in Eq. (1.41) reads

$$\delta\bar{\varepsilon}_0 = 2\,\mathrm{Re}\int_{-\infty}^{\infty}\int S(\xi, \eta)\left[F - B_0 \frac{F}{|F|}\right] \delta F^* \mathrm{d}\xi\,\mathrm{d}\eta \qquad (1.43)$$

An iterative equation that allows one to obtain the function $F(\xi, \eta)$ minimizing the functional of Eq. (1.41) follows from the requirement that the integrand in Eq. (1.43) be zero, and has the form

$$\bar{F}_{n+1} = F_n + S\left[B_0 \frac{F_n}{|F_n|} - F_n\right] = (1 - S)F_n + SB_0 \frac{F_n}{|F_n|} \tag{1.44}$$

where F_n and $\bar{F}_{n+1}$ are the functions of output in the nth iteration and input in the $(n+1)$th iteration, respectively. Given the condition in Eq. (1.44), the variation of the functional in Eq. (1.43) is understood in a special manner: $\delta F_n = F_{n+1} - F_n$ is a conventional representation of the function variation and $\delta F_n = \bar{F}_{n+1} - F_n$ is that which is employed in the present case and understood as a change-over from output to input.

The advantages of the present AA algorithm are illustrated by the following examples.

— *Example 1.1*

We compute a 1D DOE [26] which is illuminated by a plane light wave and forms in the lens focal plane an image with uniform intensity on an interval $[\xi_1, \xi_2]$.

The result of application of the GS algorithm with the replacement of Eq. (1.6) is shown in Fig. 1.3. Figures 1.3.1–1.3.5 illustrate intensity distributions $I_n(\xi)$ in the frequency plane for the iterations numbered $n = 1, 4, 7, 10$, and 13, respectively, and Fig. 1.3.6 depicts the DOE phase $\varphi(u)$ forming the intensity distribution in the 13th iteration: $I_{13}(\xi)$.

Parameters of the experiment are as follows: total number of pixels is 128, number of pixels on a DOE is 16 (with the rest of the pixels in the DOE plane equal to zero), number of pixels with uniform intensity on the given image is 32. An initial phase of the DOE was chosen to be random with variance 2π and was produced by a random number generator.

It is seen from Fig. 1.3 that the calculated DOE focuses radiation into a domain of about two diffraction spots. It is also seen that the intensity r.m.s. error d is 40% in the 13th iteration, with the energy efficiency e equal to 94.2% (Fig. 1.3.5). The parameters d and e were computed using the relations

$$d = \left[\int_{\xi_1}^{\xi_2} |I_n - e|^2 \mathrm{d}\xi\right]^{1/2} e^{-1} \tag{1.45}$$

$$e = E/\bar{E} \tag{1.46}$$

where $\bar{E}$ and E are the total energy and the energy portion coming into the domain of the desired image. Figure 1.4 shows the result of the adaptive algorithm application using the replacement in Eq. (1.23), given that $I_0 = e$. Figures 1.4.1–1.4.5 show intensity distributions for the same iteration numbers as in Fig. 1.3. Figure 1.4.6 illustrates the DOE phase in the 13th

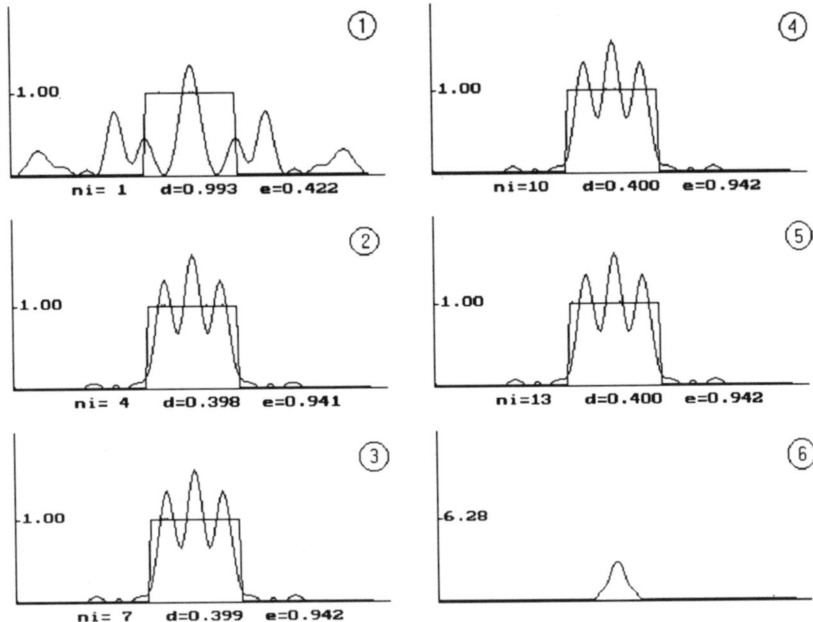

Figure 1.3 The results of the GS algorithm application: (1)–(5) the intensity distributions; (6) the DOE phase

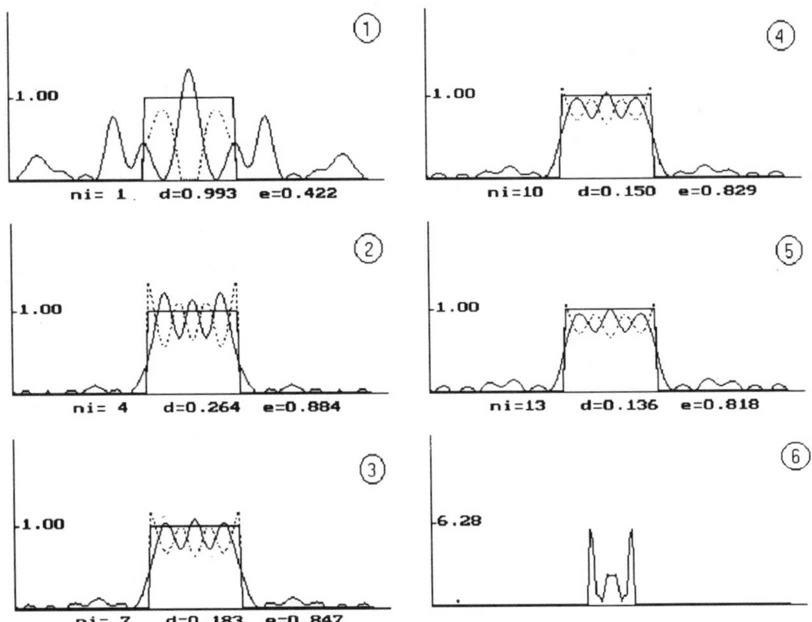

Figure 1.4 The results of the AA algorithm application: (1)–(5) the intensity distributions; (6) the DOE phase

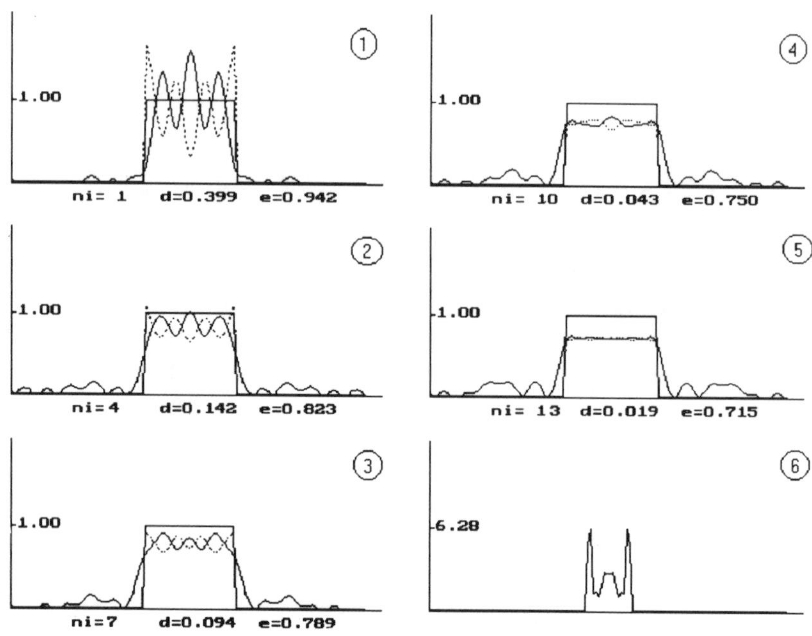

Figure 1.5 The results of collective employment of the GS and AA algorithms: (1)–(5) the intensity distributions; (6) the DOE phase

iteration. The broken line depicts the intensity $\overline{I}_n$ derived from the replacement in Eq. (1.23). It can be seen from Fig. 1.4 that in the 13th iteration the error d falls to 13.6% but the efficiency e is reduced to 81.8%. The results of collective employment of the GS and AA algorithms are shown in Fig. 1.5. The first five iterations were implemented via the replacement of Eq. (1.6) (they are omitted in Fig. 1.5) and the next 13 iterations were performed using the replacement in Eq. (1.23). Figures 1.5.1–1.5.5 depict the intensity distributions in the frequency plane computed in the iterations numbered $n = 6, 9, 12, 15, 18$, and Fig. 1.5.6 shows the DOE phase in the 18th iteration.

From the figures, the accuracy of the required intensity formation is seen to increase: the error d is reduced to 1.9%, though the efficiency e is also reduced to 71.5% (Fig. 1.5.5).

— *Example 1.2* —

We compute a DOE [75] which is illuminated by a laser light beam with the Gaussian intensity

$$A_0(u, v) = \exp\left[-\frac{u^2 + v^2}{r^2}\right] \tag{1.47}$$

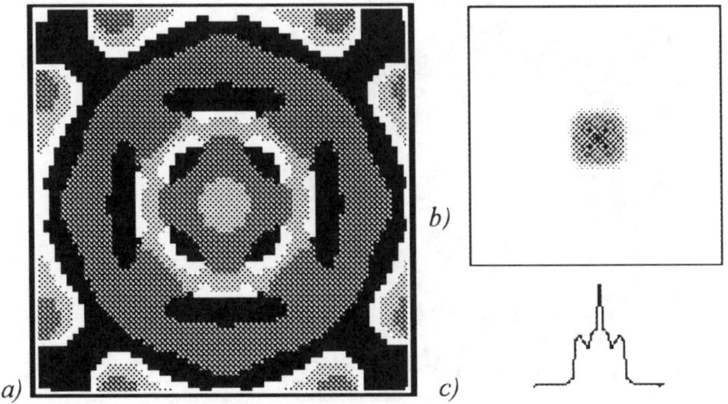

Figure 1.6 The DOE calculation using a GS algorithm: (a) the DOE phase; (b) the generated image; (c) the cross-section of the image intensity

and produces a 'soft square' image for the far field diffraction, which means that the required amplitude distribution in the observation plane is chosen to be a supergaussian function

$$B_0(\xi, \eta) = \exp\left[-\frac{\xi^{2n} + \eta^{2n}}{a^{2n}}\right], \quad n = 1, 2, \ldots \tag{1.48}$$

where $2a$ is the square effective side and r is the Gaussian beam radius.

The Fresnel transform, Eq. (1.1), and its inverse, Eq. (1.7), were calculated using a fast Fourier transform (FFT) algorithm. The pixel array was 256×256. The intensity of the illuminating Gaussian beam along the boundaries of the DOE square aperture was 10% of the maximum value at the centre. The side of the square image was about 10 Airy disks (the minimal diffraction spot). The number of iterations was 10.

Figure 1.6 shows the DOE phase modulo 2π (half-tone, 16 gradations) calculated via a GS algorithm, the image for the far field diffraction, and the horizontal cross-section of the image intensity distribution.

Figure 1.7 illustrates similar results but obtained using the AA algorithm ($\lambda = 2$): the DOE phase, the generated image, and the cross-section of the image intensity.

From Figs 1.6 and 1.7, the phases derived by these two methods are seen to be almost indistinguishable. However, a minor difference between them results in the essential difference in the generated images.

Figure 1.8 shows how the functional ε_0 value depends on the number of iterations for the case of the AA algorithm at various λ: 0.25, 0.5, 1.0, 1.25, 1.5, and 2.0. It is seen from Fig. 1.8 that the convergence rate of the iterative algorithm increases as λ increases from 0 to 2.

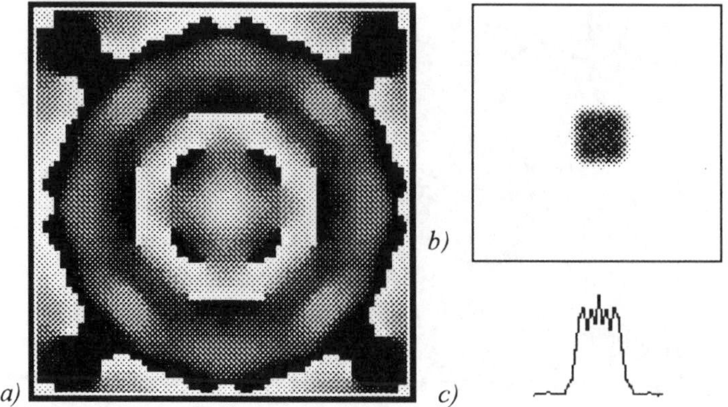

Figure 1.7 The DOE calculation using the AA algorithm: (a) the DOE phase; (b) the generated image; (c) the cross-section of the image intensity

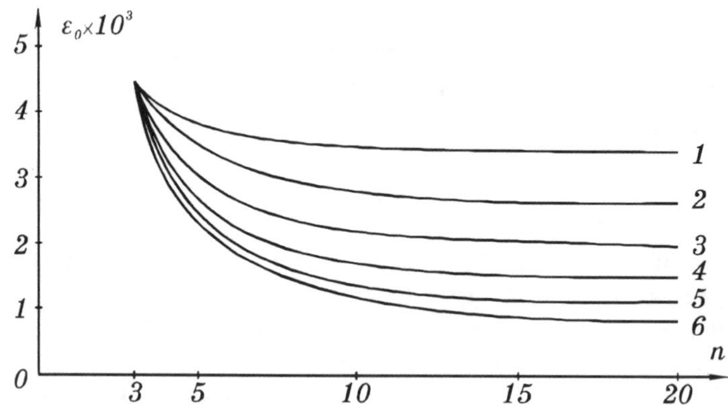

Figure 1.8 Functional ε_0 vs the number of iterations: $\lambda = 0.25$ (1), 0.5 (2), 1 (3), 1.25 (4), 1.5 (5), 2 (6)

— *Example 1.3*

We compute an optical element that can effectively locate the position [76]. This element is part of a device for determining the coordinates or the transverse displacement of the point source image. A schematic of the device is shown in Fig. 1.9.

A coherent light source S illuminates a spherical lens L behind which a DOE is placed (Fig. 1.9). The DOE is computed in such a way that the point source image has the intensity distribution in the form of a cross. Each linear component of the cross-shaped image intersects a linear photosensor as shown in Fig. 1.10. As the linear photosensor (with a characteristic length

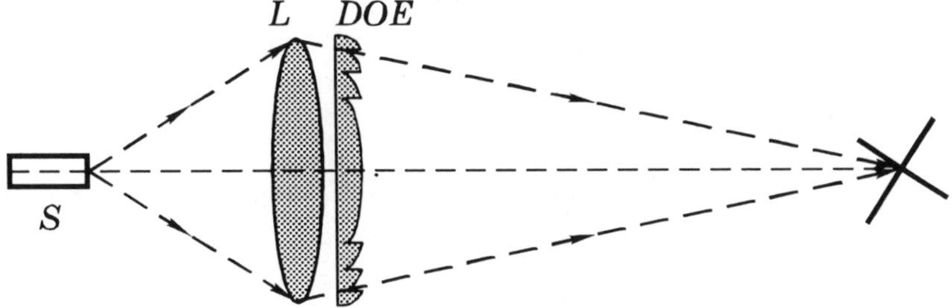

Figure 1.9 Optical setup for the generation of a cross-shaped image

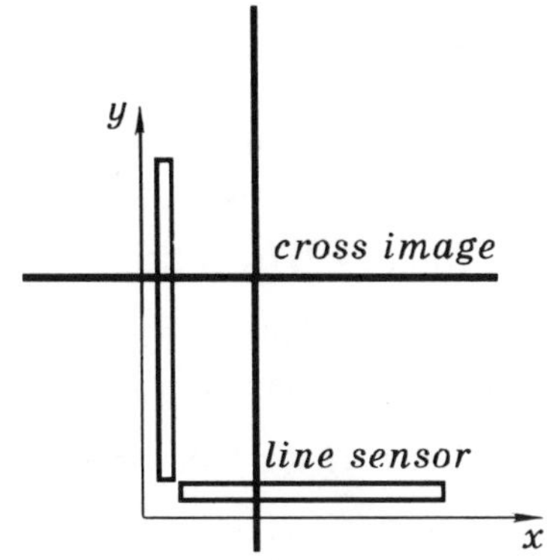

Figure 1.10 The device image plane

of input window of 10 mm) one can employ a silicon semiconducting multiscan-type structure whose output analogous signal in the form of voltage drop is proportional to the distance between the point of maximum light intensity and the sensor initial point. Characteristic parameters of such a sensor are as follows: a voltage change of 0.1 mV is equivalent to the displacement of the maximum light intensity along the sensor axis by 0.2 μm.

For computing a DOE in the present context, a paraxial relationship connecting the amplitude distribution in the source plane and the cross image was used [3]

$$F(\xi, \eta) = \frac{k^2}{ab} \exp\left[i \frac{k}{2b} (\xi^2 + \eta^2) \right] \times \int_{-\infty}^{\infty} \int W_0(u, v) \exp\left[i \frac{k}{2a} (u^2 + v^2) \right] H\left(u + \frac{b}{a} \xi, v + \frac{b}{a} \eta \right) \mathrm{d}u \, \mathrm{d}v \tag{1.49}$$

where a and b are, respectively, the distances from the source to the lens and from the lens to the image (Fig. 1.9) connected by the relation

$$\frac{1}{a}+\frac{1}{b}=\frac{1}{f} \tag{1.50}$$

f is the lens focal length, and $H(u,v)$ is the impulse response function given by

$$H(u,v)=\int_Q\int P(\xi,\eta)\,\tau(\xi,\eta)\exp\left[-i\frac{k}{b}(u\xi+v\eta)\right]\mathrm{d}\xi\mathrm{d}\eta \tag{1.51}$$

The function $P(\xi,\eta)$ in Eq. (1.51) is called the lens pupil function, and Q specifies the aperture shape of a diaphragm limiting the lens. The function $\tau(\xi,\eta)$ is the DOE transmittance amplitude being purely phase

$$\tau(\xi,\eta)=\mathrm{e}^{i\varphi(\xi,\eta)} \tag{1.52}$$

The phase $\varphi(\xi,\eta)$ of the optical element is found as the solution of the integral equation

$$I(u,v)=|H(u,v)|^2 \tag{1.53}$$

where $I(u,v)$ is the required intensity distribution for the impulse response function. For the cross, this intensity function takes the form

$$I(u,v)=\begin{cases} I_0; & -l<u<l, v=0 \text{ or } u=0, -l<v<l \\ 0; & \text{otherwise}\end{cases} \tag{1.54}$$

where I_0 is a constant value and $2l$ is the length of vertical and horizontal components of the cross.

Equation (1.53) was solved using the AA algorithm of Eq. (1.19) at $\lambda = 2$. Parameters of the calculation are as follows: the DOE was limited by a round diaphragm with a radius of 64 pixels, an array dimension of 256×256 pixels was chosen to perform the Fourier transform, each side of the cross was 170 pixels, and the width was 1 pixel. Shown in Figs 1.11a and 1.12a are the optical element phases calculated in 25 iterations with 256 and 5 gradations, respectively. The squared module of the impulse response function for a lens–DOE combination is shown in Figs 1.11b and 1.12b, for the phases with 256 and 5 gradations, respectively. The horizontal and vertical profiles for the obtained patterns of the phase and the impulse response function are also shown in Figs 1.11 and 1.12.

1.3 Adaptive–Multiplicative Algorithm

This section deals with another iterative algorithm for computing DOEs which is also able to increase the convergence rate of the GS algorithm. It

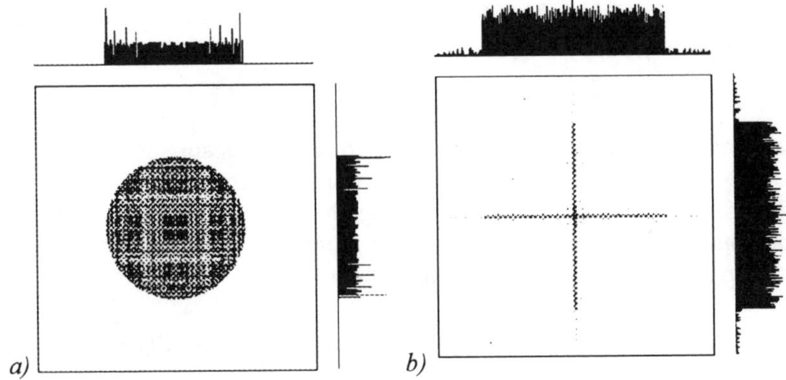

Figure 1.11 (a) the DOE phase, 256 gradations and (b) the cross in the image plane

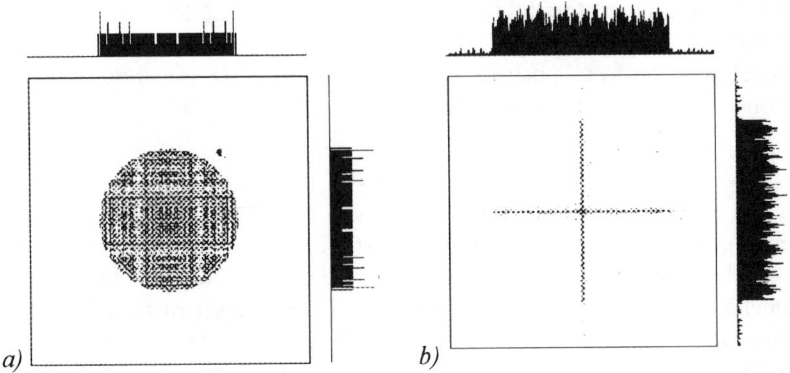

Figure 1.12 (a) the DOE phase, 5 gradations and (b) the cross in the image plane

is found that the simplest way to derive this algorithm is to use an identity which should hold in the plane of image production

$$I_0(\xi, \eta) = F(\xi, \eta)F^*(\xi, \eta) \tag{1.55}$$

Using the inverse operator $\hat{L}^{-1}$ we obtain in the DOE plane

$$W(u,v) = \hat{L}^{-1}\left[\frac{I_0(\xi, \eta)}{F^*(\xi, \eta)}\right] = \hat{L}^{-1}\left[\frac{I_0 F}{|F|^2}\right] \tag{1.56}$$

where $\hat{L}^{-1}$ is the inverse Fresnel transform.

Equation (1.56) makes it possible to suggest an iterative procedure for searching for the DOE phase (the argument of the function $W(u, v)$) which is much the same as in the GS algorithm but instead of Eq. (1.6) employs the following replacement

$$\bar{F}(\xi,\eta) = I_0(\xi,\eta)F(\xi,\eta)|F(\xi,\eta)|^{-2} \tag{1.57}$$

The drawback of Eq. (1.57) is its irregularity, because it appears to be possible for the module of $F(\xi,\eta)$ in the denominator to become zero. The regular formulation of the task of finding the DOE phase should be based on the consideration of a certain residual functional with regularization [77]. For this purpose, let us consider the functional

$$\varepsilon_2 = \int_{-\infty}^{\infty}\int [|F|^2 - I_0]^2 \mathrm{d}\xi \mathrm{d}\eta + \alpha \int_{-\infty}^{\infty}\int Q|F|^2 \mathrm{d}\xi \mathrm{d}\eta \tag{1.58}$$

The first term in Eq. (1.58) is the r.m.s. deviation of the produced intensity in the observation plane from the required intensity $I_0(\xi,\eta)$.

In contrast to the functional in Eq. (1.11), this term is a fourth-order functional relative to the light field amplitude. The second term in Eq. (1.58) is stabilizing.

Note that the real positive definite function $Q(\xi,\eta)$ should obey the Tikhonov method of regularization [78] and be given by

$$Q(\xi,\eta) = \sum_{n=0}^{N} |C_n|(\xi^2 + \eta^2)^n \tag{1.59}$$

where C_n are arbitrary coefficients. The stabilizing constant $\alpha \geq 0$ determines a lower bound of variation of the functional ε_2. Minimizing the functional ε_2 means choosing the smoothest function from the set of functions $F(\xi,\eta)$ nearest in modulus to the function $\sqrt{[I_0(\xi,\eta)]}$. An iterative procedure of minimizing the functional of Eq. (1.58) is built using interim functionals quadratic relative to the desired function $G_n(\xi,\eta)$

$$\varepsilon_{2n} = \int_{-\infty}^{\infty}\int |F_n G_n^* - I_0|^2 \mathrm{d}\xi \mathrm{d}\eta + \alpha \int_{-\infty}^{\infty}\int Q|G_n|^2 \mathrm{d}\xi \mathrm{d}\eta \tag{1.60}$$

where $F_n(\xi,\eta)$ is the complex amplitude computed in the nth iteration step in the observation plane and $G_n(\xi,\eta)$ is the function minimizing the functional ε_{2n}. A variation of the functional in Eq. (1.60) relative to the function G_n is

$$\delta\varepsilon_{2n} = 2\,\mathrm{Re}\left\{ \int_{-\infty}^{\infty}\int [|F_n|^2 G_n - I_0 F_n + \alpha Q G_n]\,\delta G_n^* \,\mathrm{d}\xi \mathrm{d}\eta \right\} \tag{1.61}$$

The minimum of the functional (1.60) is achieved provided that the variation in Eq. (1.61) becomes zero. As is seen from Eq. (1.61), this becomes possible on the condition that

$$G_n = \frac{I_0 F_n}{|F_n|^2 + \alpha Q} \tag{1.62}$$

A comparison of Eqs (1.62) and (1.57) shows that the function $G_n(\xi, \eta)$ is a regular analogue of the function $\bar{F}(\xi, \eta)$. The replacement in Eq. (1.62) minimizes the interim functionals ε_{2n} in each iteration. However, because of the convergence of the iterative procedure, the difference $|F_n - G_n|$ decreases with increasing n and the functionals ε_{2n} tend to the functional ε_2. Thus, it may be concluded that in the limit the replacement in Eq. (1.62) minimizes the original functional in Eq. (1.58). The preceding reasoning substantiates the possibility of iteratively finding the DOE phase in a way similar to the GS algorithm, but with the replacement of Eq. (1.62) used in place of the replacement of Eq. (1.6).

This algorithm is called an adaptive–multiplicative (AM) algorithm, as distinct from the AA algorithm. Its superiority to the GS algorithm can be seen from the following example.

— *Example 1.4*

We calculate a DOE [77] which is illuminated by a laser beam with the Gaussian intensity distribution and dedicated to form the image of a 'soft circle' for the far field diffraction. The required intensity in the observation plane is given by

$$I_0(\xi, \eta) = \exp\left[-2\left(\frac{\xi^2 + \eta^2}{a^2}\right)^n\right] \tag{1.63}$$

where $a/\sqrt{2}$ is the effective radius of the 'soft circle' and $n = 1, 2, 3, \ldots$ The radius a was chosen to be equal to about 20 radii of the Airy disk. The pixels array was 256×256.

Figure 1.13 depicts the DOE phase modulo 2π (16 gradations), calculated

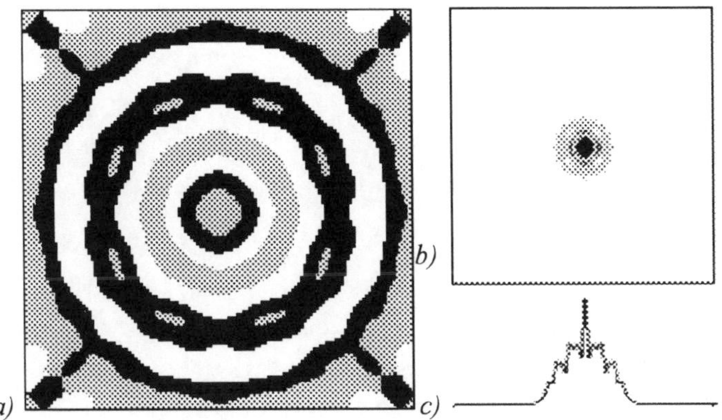

Figure 1.13 Computation of a DOE using the GS algorithm: (a) the DOE phase; (b) the diffraction pattern; (c) the intensity distribution in the central cross-section

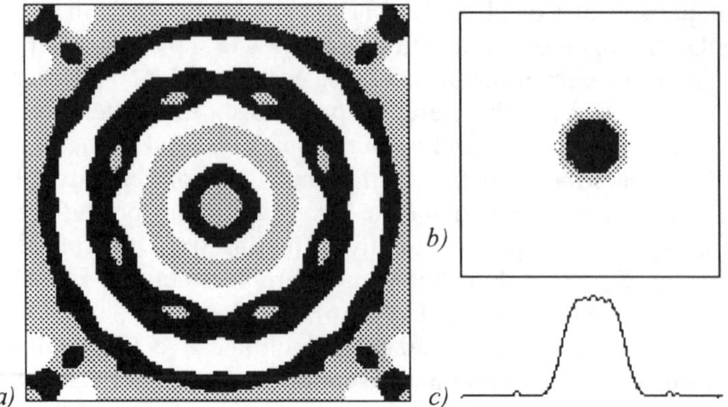

Figure 1.14 Computation of a DOE using the AM algorithm: (a) the DOE phase; (b) the diffraction pattern; (c) the intensity distribution in the central cross-section

using the GS algorithm after 20 iterations, the resulting diffraction pattern, and light intensity distribution in the central section of this pattern.

Figure 1.14 illustrates analogous results derived by the AM algorithm: the phase calculated after 20 iterations, the diffraction pattern, and the pattern of the central cross-section.

As can be seen from the comparison of these patterns, the two phases are weakly different. Nevertheless, this minor difference will cause the diffraction patterns formed to differ essentially.

Figure 1.15 shows the value of the functional ε_2 in Eq. (1.58) versus the number of iterations for the GS algorithm ($\alpha = 0$) and for the AM algorithm

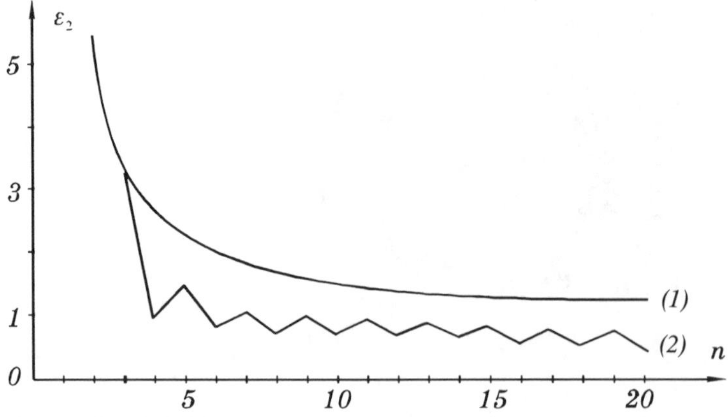

Figure 1.15 Functional ε_2 vs the number of iterations: (1) for the GS algorithm; (2) for the AM algorithm

($\alpha = 10^{-8}$). From Fig. 1.15, the adaptive algorithm is seen to have better convergence: curve 2 attains a value of 1 after 4 iterations, whereas curve 1 attains a value of 1.2 after 20 iterations.

Note, however, that Fig. 1.15 also reveals a drawback of the AM algorithm: its non-monotone convergence. It is noteworthy that if we take the values of the functional in Eq. (1.58) for every other iteration (see Fig. 1.15, curve 2), they will decrease monotonically. For all examples considered, the initial phase guess was chosen to be stochastic, and the results are almost independent of its specific realization.

Note that the lack of circular symmetry in the phase patterns shown in Figs 1.13a and 1.14a can be explained by the distortion of circular symmetry of the function $I_0(\xi, \eta)$ in Eq. (1.63) for discrete values of the variables ξ and η, in which case circles become polygons.

Note also that it is difficult to substantiate theoretically the optimal choice of the stabilizing constant α and of the polynomial power N for the function $Q(\xi, \eta)$, which affect the functional in Eq. (1.58). However, for a variety of practical applications it will suffice to choose the function $Q(\xi, \eta)$ as a monomial $(\xi^2 + \eta^2)^n$ at $n = 1, 2$, and find the constant α by fitting.

Figure 1.16 shows the r.m.s. error

$$\sigma = \left[\int_{-\infty}^{\infty}\int |I_0(\xi, \eta) - |F_n(\xi, \eta)|^2|^2 \mathrm{d}\xi \mathrm{d}\eta \right]^{1/2} \times \left[\int_{-\infty}^{\infty}\int |I_0(\xi, \eta)|^2 \mathrm{d}\xi \mathrm{d}\eta \right]^{-1/2} \tag{1.64}$$

after 10 iterations versus the stabilizing constant α (Fig. 1.16a) and the monomial power n (Fig. 1.16b). The curves have been obtained for the preceding numerical example and show that the least error σ is attained at $\alpha \approx 10^{-8}$ and $n = 1$.

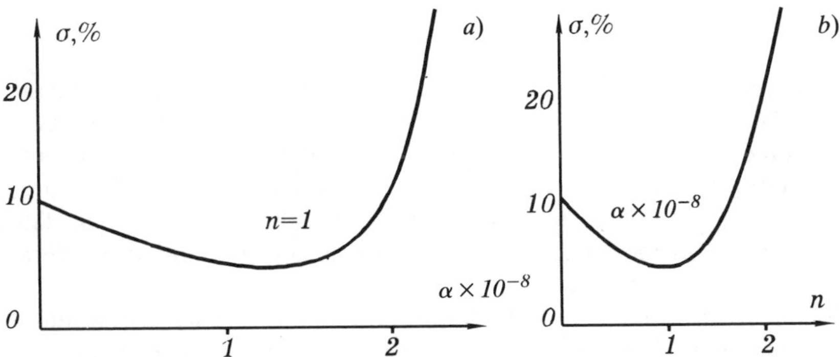

Figure 1.16 Error σ vs (a) the stabilizing constant α and (b) the power n in Eq. (1.59)

From the valuation specified by the functional in Eq. (1.58), at $\alpha = 0$, we can derive another version of the multiplicative algorithm. Consider the error functional given by

$$\varepsilon_{2p} = \int_{-\infty}^{\infty}\int [B_0^2(\xi,\eta) - |F(\xi,\eta)|^2]^p \, \mathrm{d}\xi \mathrm{d}\eta \tag{1.65}$$

The variation of this functional relative to the function $F(\xi, \eta)$ is

$$\delta\varepsilon_{2p} = 2\,\mathrm{Re}\int\int p[B_0^2(\xi,\eta) - |F(\xi,\eta)|^2]^{p-1} F(\xi,\eta)\,\delta F^*(\xi,\eta)\,\mathrm{d}\xi\mathrm{d}\eta \tag{1.66}$$

where $\delta F = \bar{F} - F$ is the difference between the input $\bar{F}$ and output F complex amplitudes in the Fourier-image plane. From the condition that the integrand in Eq. (1.66) becomes zero, an iterative equation follows for finding the input function $\bar{F}$ minimizing the functional of Eq. (1.65)

$$\bar{F}_{n+1} - F_n = -\lambda p[B_0^2 - |F_n|^2]^{p-1} F_n$$

or

$$\bar{F}_{n+1} = \{1 - \lambda p[B_0^2 - |F_n|^2]^{p-1}\} F_n \tag{1.67}$$

where λ is constant.

Equation (1.67) nonlinearly relates the output amplitude function F_n in the nth iteration to the input function $\bar{F}_{n+1}$. Note that the function F_n itself is computed using the equation

$$F_n = \mathfrak{F} D_A \mathfrak{F}^{-1}\{\bar{F}_n\} \tag{1.68}$$

where the limiting operator D_A in the DOE plane is specified by Eq. (1.16), and $\mathfrak{F}$ and $\mathfrak{F}^{-1}$ denote the Fourier transform and its inverse.

1.4 An Algorithm for Extrapolating and Interpolating the DOE Phase

The iterative algorithms dealt with in the previous sections make it possible to calculate the arrays of phase pixels of a DOE using a PC. However, it takes some time to compute a DOE of 512×512 pixels, hence computing the required arrays of $10^4 \times 10^4$ pixels will take tens of hours. Therefore, it makes sense to develop algorithms that would allow, using a small datum array of pixels of the DOE phase, the construction of large arrays without distortion of the image format.

In this section, using a unified approach, we treat such algorithms for extrapolating and interpolating DOE phase pixels which employ two $N \times N$ arrays, previously calculated via iterative procedures, and make it possible to build an $N^2 \times N^2$ array of DOE phases.

1.4.1 An Algorithm for Extrapolating Phase Pixels

A simple method for obtaining an array of DOE phases of arbitrary dimensionality has been proposed in [1]. This requires that an initial datum array of the phase pixels φ_{nm}, $n = \overline{1,N}$, $m = \overline{1,M}$ be repeated a desired number of times. Such a DOE will have the form of a 2D raster or a 2D diffraction grating with the basis function φ_{nm} taken as a period.

However, this method has an essential drawback. The image produced by such a raster DOE is disintegrated into isolated light spots whose size is in inverse proportion to the number of repetitions of the datum array of phase pixels. In the following, based on the idea proposed in [79], we discuss a method for extrapolating the phase pixels in a datum array, which appears to be free of this drawback [58].

For simplicity, consider a 1D case. Assume that it is desired to calculate the phase pixels φ_n, $n = \overline{1,N}$, of a DOE which forms a required intensity distribution with the pixels I_k in the spatial spectrum plane (in the lens focal plane). The pixels F_k of the light field in the spatial spectrum plane are known [80] to be related to the DOE transmittance $\exp[i\varphi_n]$ via the discrete Fourier transform

$$F_k = \sum_{n=1}^{N} e^{i\varphi_n} e^{-2\pi i(kn/N)}, \quad k = \overline{1,N} \tag{1.69}$$

To find φ_n, one needs to solve, using successive approximation, a nonlinear set of algebraic equations

$$I_k = |F_k|^2 = \left| \sum_{n=1}^{N} e^{i\varphi_n} e^{-2\pi i(kn/N)} \right|^2, \quad k = \overline{1,N} \tag{1.70}$$

Let us assume that the set of Eqs (1.70) is to be solved and the datum array of phase pixels φ_n, $n = \overline{1,N}$ is to be found. Next, consider a procedure of the construction of a large array of phase pixels ψ_m, $m = \overline{1,MN}$, which also forms the intensity of pixels I_k in the far diffraction field. We shall build the large array of pixels ψ_m by multiplying the datum array φ_n, with some new phase ν_m ($m = \overline{1,M}$) added

$$\psi_m = \begin{cases} \varphi_n + \nu_1, & m = \overline{1,N};\ n = \overline{1,N} \\ \varphi_n + \nu_2, & m = \overline{(N+1),2N};\ n = \overline{1,N} \\ \dots & \\ \varphi_n + \nu_M, & m = \overline{(MN-N+1),MN};\ n = \overline{1,N} \end{cases} \tag{1.71}$$

In this case, an optical element with the phase ψ_m, $m = \overline{1,MN}$, will form in the spatial spectrum plane the following complex amplitude pixels

$$\begin{aligned} R_k = \sum_{m=1}^{MN} e^{i\psi_m} e^{-2\pi i(km/MN)} &= e^{2\pi i(k/M)} \left[\sum_{n=1}^{N} e^{i\varphi_n} e^{-2\pi i(kn/MN)} \right] \\ &\times \left[\sum_{m=1}^{M} e^{i\nu_m} e^{-2\pi i(km/M)} \right] = e^{2\pi i(k/M)} P_k Q_k \end{aligned} \tag{1.72}$$

The function pixels P_k will be identical to the pregiven function pixels F_k in Eq. (1.69) at the points $k = M(n-1)+1$, $n = \overline{1,N}$

$$F_n = P_{M(n-1)+1} \tag{1.73}$$

Choosing the additional phases ν_m equal to zero, we obtain

$$Q_k = \begin{cases} 1, & k = M(n-1)+1, \quad n = \overline{1,N} \\ 0, & k \neq M(n-1)+1, \quad n = \overline{1,N} \end{cases} \tag{1.74}$$

which means that for $\nu_k = 0$ the DOE is obtained by mere repetition of the datum phase φ_n and has the form of a raster, as has been proposed in [1]. As a result, the desired image generated in the spectrum plane consists of a set of isolated light spots, i.e. we have

$$R_k = \begin{cases} F_n, & k = M(n-1)+1, \quad n = \overline{1,N} \\ 0, & k \neq M(n-1)+1, \quad n = \overline{1,N} \end{cases} \tag{1.75}$$

where $k = \overline{1,MN}$.

To avoid disintegration of the generated image into isolated points, it is necessary that the module of Q_k be equal to 1 at $k = \overline{1,M}$. Thus, we propose that the following set of algebraic equations should be iteratively solved

$$1 = \left| \sum_{k=1}^{M} \mathrm{e}^{i\nu_k} \mathrm{e}^{-2\pi i(kl/M)} \right|^2, \quad l = \overline{1,M} \tag{1.76}$$

From Eq. (1.72) it follows that the image intensity pixels $|R_k|^2$ are equal to the required intensity pixels I_k, for all $k = \overline{1,MN}$. The time taken to compute the large array of the ψ_m phase will be greatly reduced. Actually, instead of iteratively calculating the $MN \times MN$ array, we solve two sets of N and M equations. If $M = N$, the method needs computational efforts that are proportional not to M^2 but to $2M$, thus giving an $M/2$-fold gain.

It is pertinent to note that, in Eq. (1.76), instead of unity we can choose functions matched to the variation of the calculated intensity, thus facilitating a decrease in the distortions upon the multiplication of the datum array. Let us give a numerical illustration of the method considered.

— *Example 1.5* —

At a preparatory stage, we calculated a datum array of phase of 32×32 pixels for a DOE reconstructing a light square of 9×9 pixels with uniform intensity in the spatial spectrum plane. Using the GS algorithm with adaptive correction [26] we calculated the phase shown in Fig. 1.17a. The square reconstructed by such a DOE is shown in Fig. 1.17b. The relative r.m.s. deviation of the calculated intensity from the desired uniform intensity is 10% and the reconstruction efficiency is 89%. By the reconstruction efficiency we imply here the ratio of the light energy that strikes a preset spectrum domain to the total energy. Figure 1.18a illustrates the phase of a raster DOE of 256×256 pixels that has been derived from the datum phase pixels (Fig.

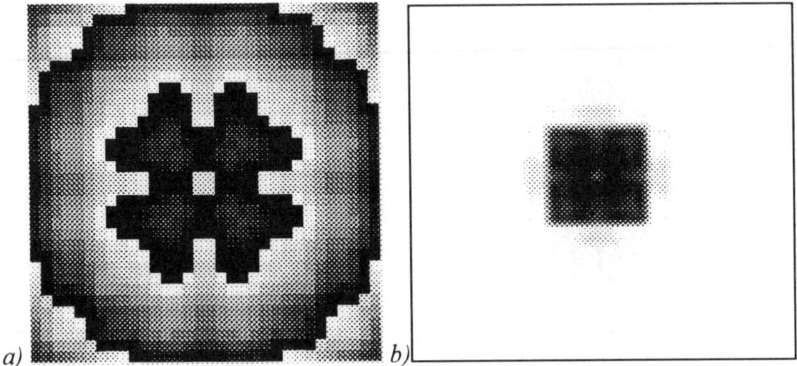

Figure 1.17 Calculation of the basic DOE: (a) the DOE phase; (b) the diffraction pattern

Figure 1.18 Extrapolation of the DOE phase without shifts: (a) the DOE phase; (b) the diffraction pattern

1.17a) by mere multiplication. Figure 1.18b illustrates the results of reconstruction. The square image is seen to disintegrate into a set of regular light spots.

Figure 1.19a depicts the phase of the raster DOE of 256×256 pixels synthesized through the multiplication of the basic phase (Fig. 1.17a) with the additional phase shifts that have been calculated upon iteratively solving the set of Eqs (1.76) of 8×8 dimension.

The square reconstructed using such a DOE is depicted in Fig. 1.19b. The reconstruction efficiency is found to change insignificantly and equals 80%, with the r.m.s. deviation increasing to 16%. Such an increase in error is due to the enlarged square size (8 times), which means that on the square additional points appeared and have been used for estimating the error. An increase in the size of the reconstructed image (8 times) is a drawback of the extrapolating method. This prompted us to consider in the following

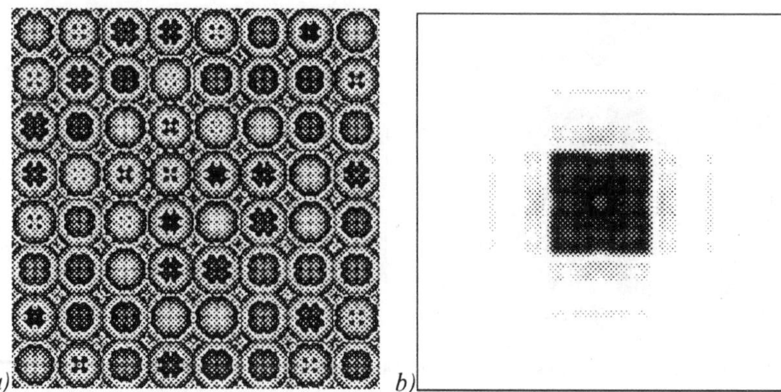

Figure 1.19 Extrapolation of the DOE phase with shifts: (a) the DOE phase; (b) the diffraction pattern

section an interpolating method that does not lead to the increase of the image.

1.4.2 An Algorithm for Interpolating the Phase Pixels

There is a class of optical elements called focusators [37] that, just as DOEs, are phase-only elements but, taken in combination with a spherical lens, can focus the laser light into small areas of the spatial spectrum of several Airy disk diameters. For calculating such focusators the extrapolation method appears to be unsuitable, because it leads to the increase of the size of the diffraction pattern (image) formed, thus reducing the light energy density.

Below is the algorithm for interpolating DOE phase pixels that leads to the increased dimension of the pixels array but does not lead to increased image size.

Assume that there exists a calculated datum array of the phase φ_n, $n = \overline{1, N}$, that obeys the set of Eqs (1.70). Let us construct a large array of the phase pixels ψ_m, $m = \overline{1, MN}$ using additional phase shifts ν_k, $k = \overline{1, M}$ using the rule

$$\psi_m = \begin{cases} \varphi_1 + \nu_k, & m = \overline{1, M};\ k = \overline{1, M} \\ \varphi_2 + \nu_k, & m = \overline{(M+1), 2M};\ k = \overline{1, M} \\ \vdots & \\ \varphi_{l+1} + \nu_k, & m = \overline{(lM+1), (l+1)M};\ k = \overline{1, M} \\ \vdots & \\ \varphi_N + \nu_k, & m = \overline{(MN-N+1), NM};\ k = \overline{1, M} \end{cases} \tag{1.77}$$

according to which the large array of the ψ_m phase can be obtained if we repeat M times each pixel φ_n of the datum array, with the pixels ν_k added.

Hence, the dimension of the large array will be M^2 times greater than that of the datum array.

The kinoform with the phase ψ_m will form in the lens focal plane the intensity pixels that can be represented via the discrete Fourier transform as follows

$$R_k = \sum_{m=1}^{MN} e^{i\psi_m} e^{-(i2\pi/MN)km} = e^{i2\pi k/N} \left[\sum_{n=1}^{N} e^{i\varphi_n} e^{-(i2\pi/N)kn} \right] \times \left[\sum_{l=1}^{M} e^{i\nu_l} e^{-(i2\pi/NM)kl} \right] = \bar{P}_k \bar{Q}_k e^{i2\pi k/N} \quad (1.78)$$

The function pixels $\bar{P}_k$ will coincide with the pixels F_k of the desired function in Eq. (1.69) not only for $k = \overline{1, N}$, but also for $k = \overline{(N+1), 2N}$, $k = \overline{(2N+1), 3N}$ and so on. This means that the kinoform whose phase is specified by Eq. (1.77) will form the needed image in M diffraction orders, with their amplitudes changing from order to order in accordance with the second multiplier $\bar{Q}_k$ in Eq. (1.78). However, the light energy distribution in diffraction orders can be controlled by an appropriate choice of the additional phase shifts ν_k. For example, if we need to concentrate the entire light energy in the zero diffraction order, where the desired image F_k is generated, we can choose all the additional phases to be zero: $\nu_k = 0$, $k = \overline{1, M}$. The second multiplier in Eq. (1.78) will then take the form

$$\bar{Q}_k = \operatorname{sinc}\frac{\pi k}{N}, \quad k = \overline{1, MN} \quad (1.79)$$

This will be the function of light distribution in the diffraction orders. The maximum of this function will coincide with the zero diffraction order. However, a drawback of the function in Eq. (1.79) shows up in the deterioration of the required function pixels I_k, $k = \overline{1, N}$, for the values of k close to N directly in the zeroth order. An ideal function that makes the light intensity zero in all orders except zero and attains a constant value in the zero order is the rect-function

$$\bar{Q}_k = \operatorname{rect}(k/N) = \begin{cases} 1, & k < N \\ 0, & k > N \end{cases} \quad (1.80)$$

In this case, the additional pixels ν_k should be sought as the solution to the set of algebraic equations similar to Eq. (1.76)

$$\operatorname{rect}(k/N) = \left| \sum_{l=1}^{M} e^{i\nu_l} e^{-(2\pi/MN)kl} \right|^2, \quad k = \overline{1, MN} \quad (1.81)$$

The solution to Eqs (1.81) can be derived via iterations, as in [14,26]. Therefore, we have shown that to build the $MN \times MN$ interpolated array of the DOE phase through Eq. (1.77), one should iteratively solve Eq. (1.70) and find an $N \times N$ datum array of phases, and solve Eq. (1.81) and find an $M \times M$ array of the additional phase shifts.

If one needs to form an image localized in the range of low spatial frequencies of the lens focal plane (such images are produced by the focusator), one can employ additional pixels with zero values, i.e. generate the hologram phase as a raster.

The following example illustrates an operation of the interpolation algorithm.

— *Example 1.6* —

The initial phase of the DOE (32×32 pixels) was chosen as shown in Fig. 1.17a. The DOE with such a phase should form in the lens focal plane a light square of 9×9 pixels with uniform intensity (see Fig. 1.17b). The r.m.s. deviation of the intensity of the generated square (Fig. 1.17b) from a required constant value was 10%, with an 89% energy efficiency.

Figure 1.20a depicts an interpolated DOE phase (256×256 pixels) which is derived from the basic phase (Fig. 1.17a) by means of an 8-fold replication of each pixel of the basic phase in vertical and horizontal directions. This procedure corresponds to the algorithm of interpolation using additional phase shifts, provided that these shifts are chosen to be zero.

Figure 1.20b shows the light square (9×9 pixels) which is formed by the DOE with such a phase (Fig. 1.20a). In this case the r.m.s. error is seen to be somewhat larger (13%), with lower efficiency (85%). However, these variations proved to be inessential in practice.

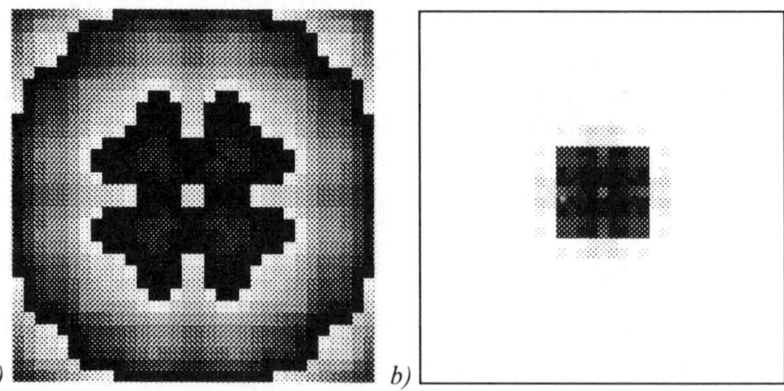

Figure 1.20 Phase interpolation without shifts: (a) the DOE phase; (b) the diffraction pattern

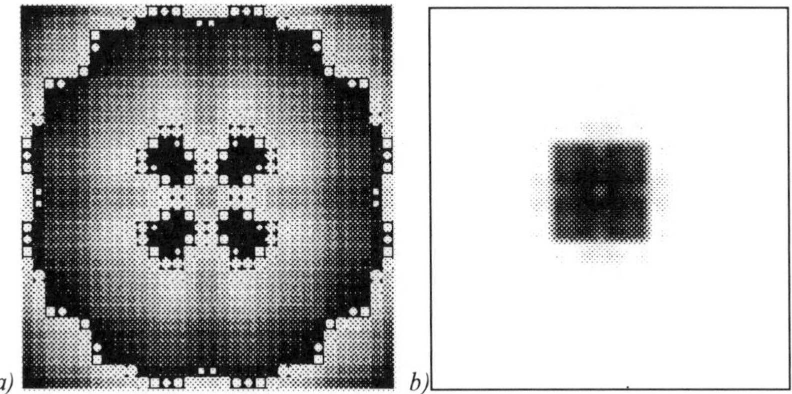

Figure 1.21 Phase interpolation with shifts: (a) the DOE phase; (b) the diffraction pattern

Figure 1.21a depicts an interpolated DOE phase (256×256 pixels) that has also been obtained by an 8-fold replication of each pixel but with additional phase shifts which are no longer zero and are found by iteratively solving the set (1.81). The light square produced by the DOE with such a phase is shown in Fig. 1.21b.

The r.m.s. deviation of the intensity from the constant amounts to 10%, with the diffraction efficiency reduced to 79%. Thus, from the example above the algorithm for interpolating hologram pixels is seen to be operative: a 64-fold increase in the number of pixels (from 32×32 to 256×256) resulted in the same deviation of the reconstructed image from the desired one as is the case for the basic DOE, which means that the developed approach to the interpolation does not cause an essential distortion of the image under synthesis.

2

Iterative Algorithms for Calculating DOEs Forming Radially Symmetrical Images

For computing optical elements dedicated to producing in a certain spatial region an intensity distribution with circular symmetry, it is useful to employ polar coordinates

$$\begin{cases} u = r\cos\varphi \\ v = r\sin\varphi \end{cases} \quad \begin{cases} \xi = \rho\cos\theta \\ \eta = \rho\sin\theta \end{cases} \tag{2.1}$$

In this case, the Fresnel integral transform, Eq. (1.1), in polar coordinates takes the form

$$F(\rho,\theta) = \frac{ik}{2\pi z}\mathrm{e}^{ikz}\int_0^{\infty}\int_0^{2\pi} W(r,\varphi)\exp\left[\frac{ik}{2z}\{r^2+\rho^2-2r\rho\cos(\varphi-\theta)\}\right] \times r\,\mathrm{d}r\,\mathrm{d}\varphi \tag{2.2}$$

In order for the module of the complex amplitude $F(\rho,\theta)$ in the observation plane to be independent of the azimuth angle θ, we confine consideration to the set of functions given by

$$W(r,\varphi) = g(r)\mathrm{e}^{im\varphi},\ m = 1,2,3,\ldots \tag{2.3}$$

Instead of Eq. (2.2) we derive

$$F(\rho,\theta) = i^{m+1}kz^{-1}\mathrm{e}^{ikz}\mathrm{e}^{im\theta}\int_0^{\infty} g(r)\mathrm{e}^{(ik/2z)(r^2+\rho^2)}J_m\left(\frac{kr\rho}{z}\right)r\,\mathrm{d}r \tag{2.4}$$

where

$$J_m(x) = \frac{(-i)^m}{2\pi}\int_0^{2\pi}\mathrm{e}^{i(mt+x\cos t)}\,\mathrm{d}t \tag{2.5}$$

is the Bessel function of the first kind and mth order.

From Eq. (2.4) it follows that to search for the DOE phase within the framework of the scalar theory of Fresnel's diffraction, one needs to solve the integral equation

$$I_0(\rho) = |F(\rho, \theta)|^2 = \left| kz^{-1} \int_0^R A_0(r) \mathrm{e}^{i\psi(r)} \mathrm{e}^{ikr^2/2z} J_m\left(\frac{kr\rho}{z}\right) r\,\mathrm{d}r \right|^2 \tag{2.6}$$

where $I_0(\rho)$ is the radial intensity distribution for the image to be produced at a distance z from the DOE plane, $A_0(r)$ is the radial distribution of the amplitude of the illuminating beam, and $\psi(r)$ is the DOE phase sought. In the following we consider iterative methods for solving Eq. (2.6) for different situations: the production of the desired image in the spatial spectrum plane and the formation of axial light beams with a required axial intensity.

2.1 Calculation of DOEs Forming Radially Symmetric Diffraction Patterns

From Eq. (2.6) it follows that to search for the phase of a DOE iteratively, one needs to compute the Hankel transform (HT) and its inverse

$$H(\rho) = \int_0^\infty h(r) J_m(r\rho) r\,\mathrm{d}r \tag{2.7}$$

$$h(r) = \int_0^\infty H(\rho) J_m(r\rho) \rho\,\mathrm{d}\rho \tag{2.8}$$

To compute quickly the HT of Eqs (2.7) and (2.8), one can employ a method of exponential replacement of variables [81]. This method suggests that by using an exponential replacement of variables the HT is reduced to a convolution which can be calculated via the Fourier transform. Actually, after the replacement of variables

$$r = r_0 \mathrm{e}^x, \quad \rho = \rho_0 \mathrm{e}^y \tag{2.9}$$

where r_0 and ρ_0 are constant, instead of Eq. (2.7) we find

$$\bar{H}(y) = r_0^2 \int_{-\infty}^\infty \bar{h}(x) S(x+y) \mathrm{e}^{2x} \mathrm{d}x \tag{2.10}$$

where

$$\begin{aligned} \bar{h}(x) &= h(r_0 \mathrm{e}^x) \\ S(x+y) &= J_m(r_0 \rho_0 \mathrm{e}^{x+y}) \\ \bar{H}(y) &= H(\rho_0 \mathrm{e}^y) \end{aligned} \tag{2.11}$$

In order for the $S(x)$ function to tend to zero at $x \to \pm\infty$, it can be multiplied by $\exp(x/4)$ and, lest the integrand in Eq. (2.10) changes, one should also multiply the function $h(x)$ by $\exp(-x/4)$.

Because the transmission function $h(r)$ of the DOE is limited by the aperture $r \in [0, R]$, where R is the aperture radius, no problems arise for the function $\bar{h}(x)$ at $x \to \infty$.

So, after redesignations we obtain

$$h_1(x) = \bar{h}(x) r_0^2 e^{7x/4}$$

$$S_1(x+y) = S(x+y) \exp\left(\frac{x+y}{4}\right) \tag{2.12}$$

$$H_1(y) = \bar{H}(y) e^{y/4}$$

and instead of Eq. (2.10) we derive the convolution

$$H_1(y) = \int_{-\infty}^{a} h_1(x) S_1(x+y) \, \mathrm{d}x \tag{2.13}$$

$$a = \ln \frac{R}{r_0}$$

The integral in Eq. (2.13) can be expressed via the Fourier transform as

$$H(\rho) = \left[\frac{\rho}{\rho_0}\right]^{-1/4} \int_{-\infty}^{\infty} P(-w) U(w) e^{iw \ln(\rho/\rho_0)} \mathrm{d}w \tag{2.14}$$

where $P(w)$ is a Fourier image of the function $h_1(x)$ and $U(w)$ is a Fourier image of $S_1(y)$. In a similar way, we can represent the inverse of the HT, Eq. (2.8). Note that there are other methods for quickly calculating the HT [82,83].

Turning back to solving Eq. (2.6), we unite the two exponents in one function which is to be found

$$\bar{\psi}(r) = \psi(r) + \frac{k}{2z} r^2 \tag{2.15}$$

The procedure of iteratively solving Eq. (2.6) proceeds similarly to the Gerchberg–Saxton algorithm and successively employs a pair of the HT and its inverse, with limiting replacements made both in the optical element plane and in the observation plane. In the nth iteration step, the $g_n(r)$ function calculated in the DOE plane is replaced by the $\bar{g}_n(r)$ function according to

$$\bar{g}_n(r) = \begin{cases} A_0(r) g_n(r) |g_n(r)|^{-1}, & r \in [0, R] \\ 0, & r \in [0, R] \end{cases} \tag{2.16}$$

while the calculated complex amplitude $F_n(\rho)$ in the observation plane is replaced by $\bar{F}_n(\rho)$, using the relation

$$\bar{F}_n(\rho) = \sqrt{[I_0(\rho)]} F_n(\rho) |F_n(\rho)|^{-1} \tag{2.17}$$

There is no essential difference between the above replacements, Eqs (2.16) and (2.17), and the corresponding replacements for the 2D GS algorithm, Eqs (1.6) and (1.8). To enhance the convergence rate of the algorithm, instead of Eq. (2.17) one can use the replacements of the AA algorithm, Eqs (1.19), with the corresponding change-over from the 2D to the 1D radial case.

The algorithm of Eqs (1.6) and (1.8) enables the calculation of DOEs dedicated to forming, in the lens focal plane or in the far diffraction field, an image, such as a circle, a ring, or a set of rings [28,84].

— *Example 2.1* —

Figure 2.1 depicts the phase (after 22 iterations) of a DOE which is able to transform the laser light with Gaussian intensity distribution into a circle of uniform intensity in the lens focus. The intensity of the Gaussian beam on the boundaries of the DOE aperture is 0.1 of its maximum in the beam centre. The DOE radius is $R = 0.4$ mm, the number of pixels along the radius is 256, $k/f = 100\ \text{mm}^{-2}$, f is the lens focal length (in Eqs (2.6) and (2.15) one should take f for z), the circle radius in the image plane is 0.3 mm and equals approximately three radii of the minimal diffraction spot (Airy disks). Figure 2.1b illustrates the radial section of the DOE phase.

Figure 2.2 shows (a) the diffraction pattern produced in the focal plane by the DOE with such a phase and (b) its radial section. The r.m.s. deviation of the generated intensity over a preset circle from a constant value is 6% and the efficiency of focusing into the preset circle is 91%. In our case the calculation has been performed using the HT (for $m = 0$, the Fourier–Bessel transform).

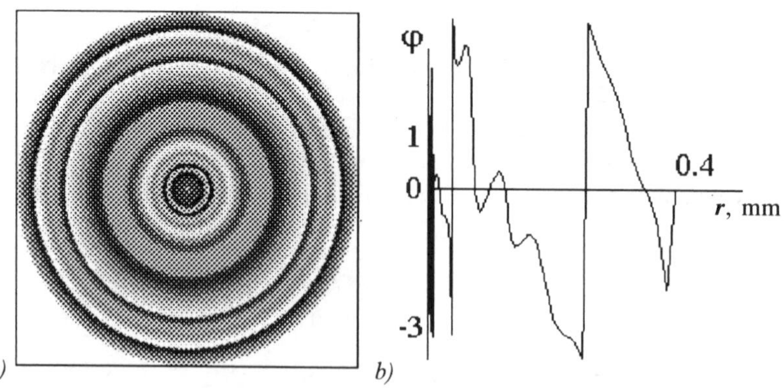

Figure 2.1 Phase of a DOE focused into a circle: (a) the DOE phase; (b) its radial section

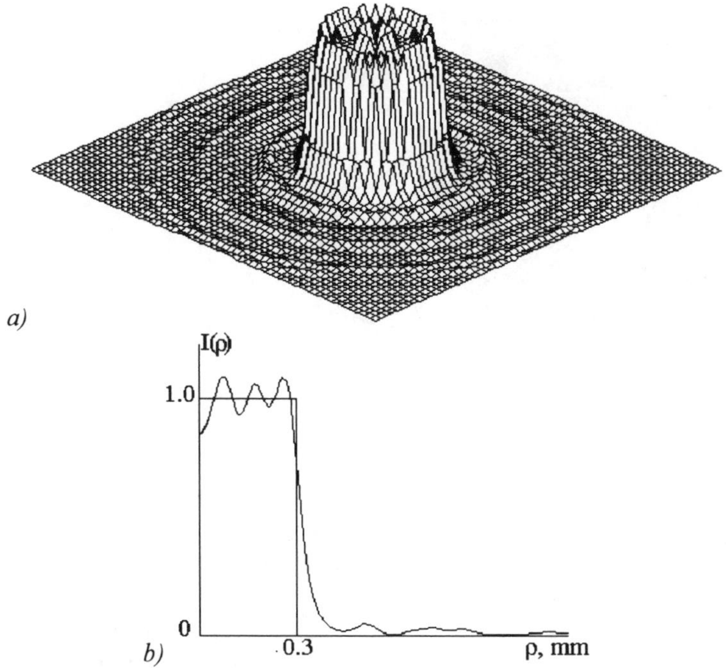

Figure 2.2 Generated image: (a) the diffraction pattern; (b) its radial section

If we need to compute a DOE with radial phase forming a ring-like intensity distribution, and if for such a calculation we wish to use the HT of zero order, there will be (almost without exception) an intensity peak in the centre of the ring. This is illustrated by the following example.

— *Example 2.2* ———

Figure 2.3 depicts the DOE radial phase calculated after 50 iteration steps at $m = 0$ and the diffraction pattern in the focal plane. The intensity is seen to have a local maximum in the centre.

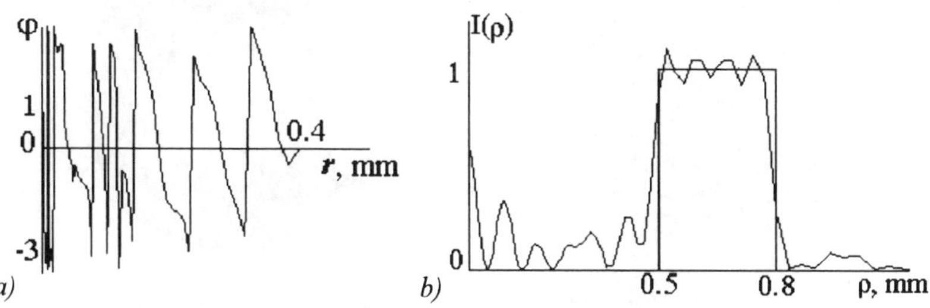

Figure 2.3 A radial DOE focusing into a ring ($m = 0$): (a) the radial DOE's phase; (b) the radial section of intensity in the focal plane

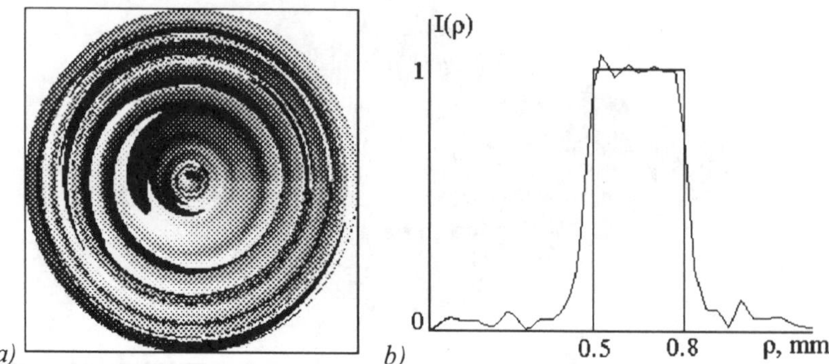

Figure 2.4 A non-radial DOE focusing into a ring ($m = 5$): (a) the DOE phase; (b) the radial section of intensity in the focal plane

If the calculation is conducted using, for example, the HT of the fifth order ($m = 5$), and the phase is represented in the form of Eq. (2.3), we will come to the result shown in Fig. 2.4: the DOE phase (not radial) and the diffraction pattern in which the central intensity is close to zero. The inside and outside radii of the ring in Figs 2.3 and 2.4 are, respectively, 0.5 mm and 0.8 mm, whereas the remaining parameters are taken to be equivalent to the previous case.

The intensity of focusing into a ring is 82% (Fig. 2.3b) and 86% (Fig. 2.4b), and the r.m.s. deviation is, respectively, 6% and 4%.

— Example 2.3

Figure 2.5 shows the result of a similar calculation of a DOE into a wide ring with uniform intensity but with different parameters of calculation. In

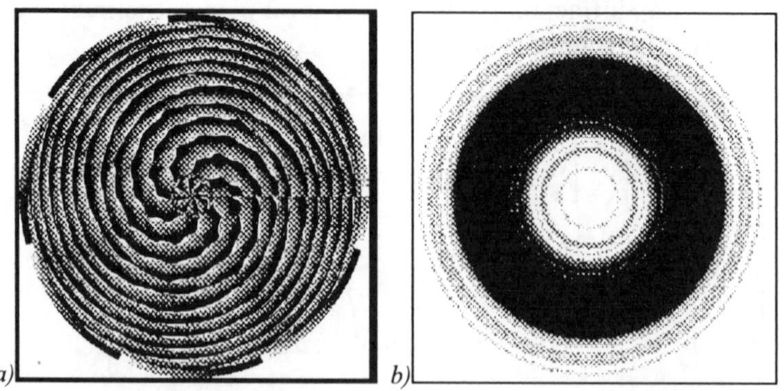

Figure 2.5 (a) Phase of a DOE focused into a ring; (b) the diffraction pattern at $m = 8$

this case we carried out the calculation via the HT of the 8th order, the DOE radius is $R = 5$ mm, the ring radii are $r_1 = 0.1$ mm and $r_2 = 0.2$ mm, and the number of pixels along the radius is 256. The efficiency of focusing into a ring is 83%; the r.m.s. deviation is 8%.

Note that the azimuth multiplier in Eq. (2.3) in the form $e^{im\varphi}$ is of significant interest as a phase optical element, using which one can perform a variety of optical transformations: the HT of the mth order [63], the production of Bessel beams of higher orders [50], the optical differentiation of radially symmetric light beams [85], and a radial analogue of the Hilbert transform [64,86].

2.2 Calculation of DOEs–Axicons

Phase optical elements able to form, in combination with a spherical lens, a light field with enhanced focal depth are called focusators into a longitudinal line-segment [33] and can be considered as general axicons [87]. They have found wide practical use: for studying the laser discharge in the gas [69], for reading data in and out of optical disks [88], for testing surfaces [89], etc.

In the subsequent text, we consider iterative algorithms for calculating the DOEs (generalized axicons) that are able to produce light beams of a pregiven intensity distribution along their central axes.

Let us assume that we need to compute the radial phase function of an optical element that forms an axial light segment bounded by the points z_1 and z_2 and characterized by a preset axial intensity distribution $I_0(z)$, $z \in [z_1, z_2]$. In this case, Eq. (2.4) will take the form

$$F(\rho, z) = \frac{k}{z} e^{ikz} e^{ik\rho^2/2z} \int_0^R \exp\left[i\varphi(r) + i\frac{kr^2}{2z} \right] J_0\left(\frac{kr\rho}{z}\right) r\,dr \qquad (2.18)$$

where R is the radius of the DOE, $\varphi(r)$ is the desired phase, and $J_0(x)$ is the Bessel function of zero order. In Eq. (2.18), the optical element is assumed to be illuminated by the plane light wave.

Since we know the intensity along the z-axis, we can put $\rho = 0$ in Eq. (2.18) and obtain (omitting the trivial cofactor $\exp(ikz)$)

$$F(0, z) = \frac{k}{z} \int_0^R \exp\left[i\varphi(r) + i\frac{kr^2}{2z} \right] r\,dr \qquad (2.19)$$

On renaming

$$\xi = \frac{k}{z}, \quad x = \frac{r^2}{2} \qquad (2.20)$$

Eq. (2.19) can be rewritten as

$$\frac{1}{\xi} F(\xi) = \int_0^a e^{i\varphi(x) + ix\xi} dx, \quad a = \frac{R^2}{2} \qquad (2.21)$$

from which the desired function of the DOE–axicon transmittance $e^{i\varphi(\rho)}$ is related to the axial complex light amplitude $F(\xi)$ via a 1D Fourier transform. Therefore, the problem has been reduced to solving the 1D integral equation

$$\frac{1}{\xi^2} I_0(\xi) = \left| \int_0^a e^{i\varphi(x)+ix\xi} dx \right|^2 \tag{2.22}$$

in which case all the algorithms proposed in Chapter 1 are directly suitable.

If we wish to calculate an axicon as an addition to a spherical lens, the phase of Eq. (2.19) should be given by

$$\varphi(r) = \varphi_0(r) - \frac{kr^2}{2f} \tag{2.23}$$

where f is the focal length of the spherical lens. In that case instead of Eq. (2.19) we can write

$$F(0,z) = \frac{k}{z} \int_0^R \exp\left[i\varphi_0(r) + i\frac{kr^2}{2}\left(\frac{1}{z} - \frac{1}{f}\right)\right] r\,dr \tag{2.24}$$

which, relative to $\varphi_0(r)$, will be equivalent to Eq. (2.21) if we use the replacement of variables

$$\xi = k\left(\frac{1}{z} - \frac{1}{f}\right), \quad x = \frac{r^2}{2} \tag{2.25}$$

To form short light segments (provided that $2d = z_2 - z_1 \ll f$, where $2d$ is the depth of focus of the lens, or the length of the preset light segment), instead of Eq. (2.24) one can use a more simple equation

$$F(\Delta z) = \frac{k}{f} \int_0^R \exp\left[i\varphi_0(r) + i\frac{k\Delta z}{2f^2} r^2 \right] r\,dr \tag{2.26}$$

where $\Delta z \in [-d, d]$ and $f = (z_1 + z_2)/2$.

On the replacement of variables

$$\xi = \frac{k\Delta z}{f^2}, \quad x = \frac{r^2}{2} \tag{2.27}$$

we finally derive the equation for searching for the DOE phase

$$I_0(\xi) = \left| \frac{k}{f} \int_0^a e^{i\varphi_0(x)+ix\xi} dx \right|^2 \tag{2.28}$$

which admits of a solution via an iterative procedure [29,48,90], with the degree of proximity of the estimator of intensity $I_n(\xi) = |F_n(\xi)|^2$ to the desired function $I_0(\xi)$ to be determined from the r.m.s. deviation

$$\delta = \left[\int_{\xi_1}^{\xi_2} |I_n(\xi) - I_0(\xi)|^2 d\xi \right]^{1/2} \left[\int_{\xi_1}^{\xi_2} I_0^2(\xi)\, d\xi \right]^{-1/2} \tag{2.29}$$

The efficiency of production of the axial light segment can be specified as

$$E_k = \int_0^{\rho_0} I(\rho, z_k)\rho \mathrm{d}\rho \bigg/ \int_0^{\infty} I(\rho, z_k)\rho \mathrm{d}\rho \tag{2.30}$$

where $I(\rho, z_k)$ is the calculated intensity in the plane which is z_k apart from the DOE and ρ_0 is the radius of the first local intensity minimum $I(\rho, z_k)$. The E_k function determines the part of light energy that goes to form the minimal diffraction spot in a given observation plane.

Note that if the desired intensity $I_0(z)$ is given by

$$I_0(z) = I_0 \sum_{n=1}^{N} \delta(z - z_n) \tag{2.31}$$

where $\delta(x)$ is the Dirac delta, the solution for Eq. (2.28), $\varphi(r)$, will be the phase of an N-focus lens with N longitudinal focii of equal intensity I_0.

— *Example 2.4*

Figure 2.6a depicts the DOE phase calculated within 100 iterations using a GS algorithm for solving Eq. (2.28). This DOE produces a light beam with the axial intensity distribution shaped as a rectangular impulse (Fig. 2.6b). Figure 2.6c illustrates the same phase in the form of an amplitude mask.

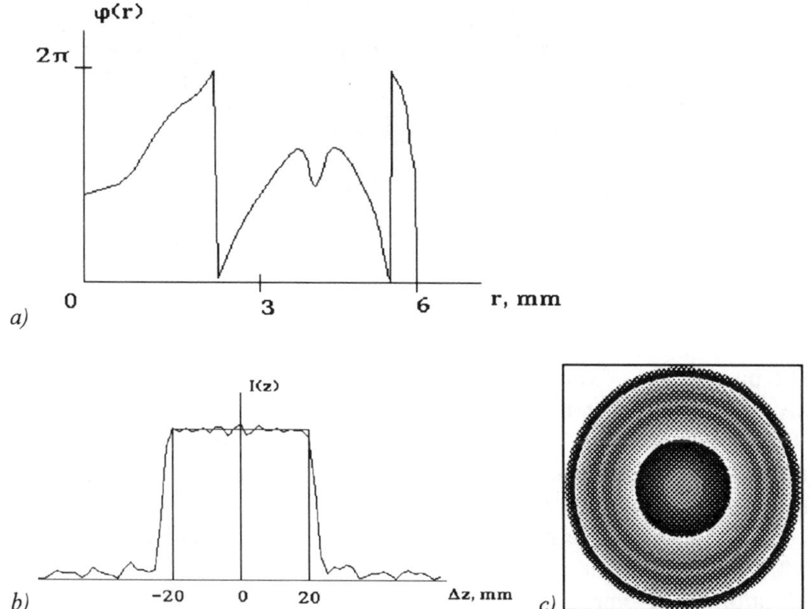

Figure 2.6 The DOE focusing into an axial line segment: (a) the radial section of the DOE phase; (b) the axial intensity; (c) the 2D DOE phase

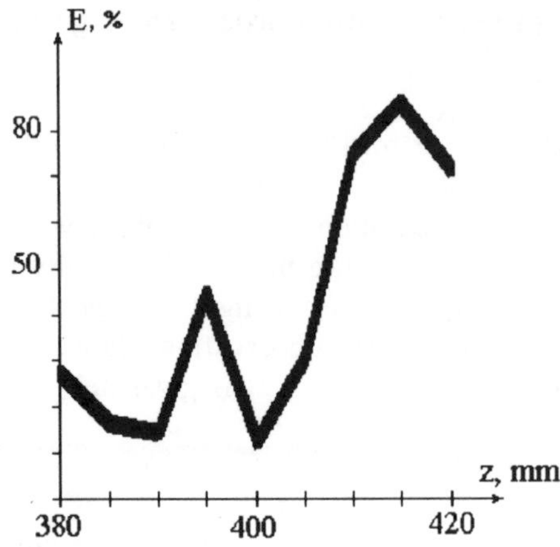

Figure 2.7 Efficiency along the line

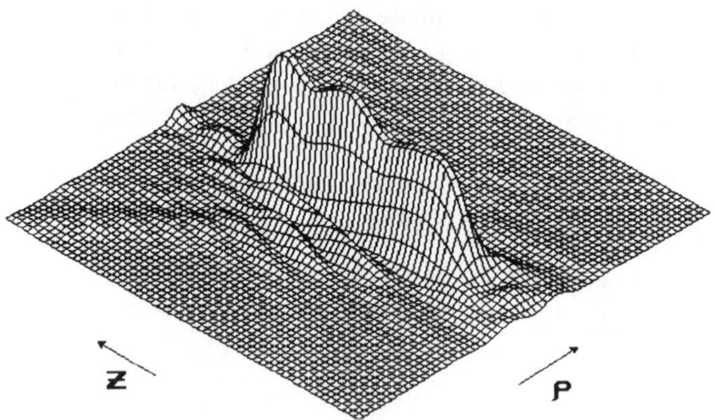

Figure 2.8 Intensity along the axis

The parameters of calculation are as follows: $N = 256$ is the total number of pixels, $R = 6$ mm is the DOE radius, $f = 400$ mm is the focal length of the lens which contains in the vicinity of its focal plane the produced light segment $2a = 40$ mm in length. The wavelength is 0.63 μm; the illuminating light beam is assumed to be plane. The error (Fig. 2.6b) is 2%, whereas the efficiency of focusing varies along the segment in a complicated manner (see Fig. 2.7) and lies in the range from 15% to 85%. A 3D plot of the intensity distribution $I(\rho, z)$ generated by the DOE with the phase shown in Fig. 2.6c is presented in Fig. 2.8. The length of the line is about four Fresnel lengths.

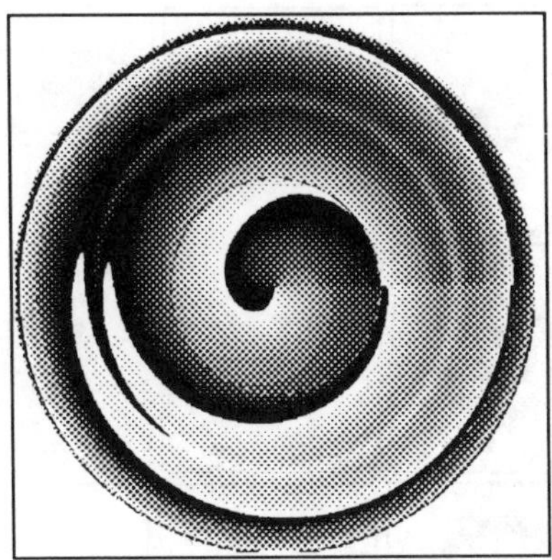

Figure 2.9 Phase of a kinoform for synthesizing a light pipe

— *Example 2.5*

A phase rotor filter, Eq. (2.3), can be used to form an axial light 'pipe', i.e. a light beam with zero axial intensity. For this purpose, a 'rotor', or 'helical', term $m\theta$ (θ is the polar angle) should be added to the phase $\varphi_0(r)$ calculated using Eq. (2.28). Figure 2.9 shows such a phase derived by adding the rotor term ($m = 1$) to the phase depicted in Fig. 2.6c.

— *Example 2.6*

For comparison, Fig. 2.10 shows diffraction patterns formed at different distances by the lens without a DOE, by the lens with a DOE having the phase $\varphi_0(r)$ (Fig. 2.6c), and by the lens with a DOE having the phase $\varphi_0(r) + \theta$ (Fig. 2.9).

— *Example 2.7*

Based on Eqs (2.28) and (2.31), we can deduce the phase of a multifocus lens.

Figure 2.11a depicts the radial distribution of phase of the DOE that forms ten longitudinal foci of equal intensity (Fig. 2.11b).

2.3 Calculation of Radially Symmetric DOEs with Quantized Phase

In this section we consider an iterative algorithm that enables us to compute radially symmetric DOEs with a small number of phase levels, thus offering

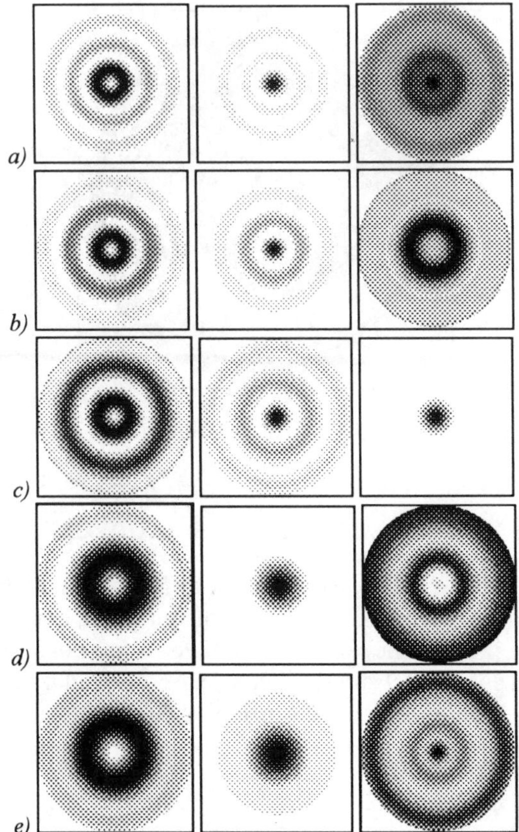

Figure 2.10 Axial intensity distribution: for the lens without a DOE (right), for the lens with the radial DOE (Fig. 2.6c) (centre), and for the lens with the rotor DOE (Fig. 2.9) (left): $z = 380$ mm (a), 390 mm (b), 400 mm (c), 410 mm (d), 420 mm (e)

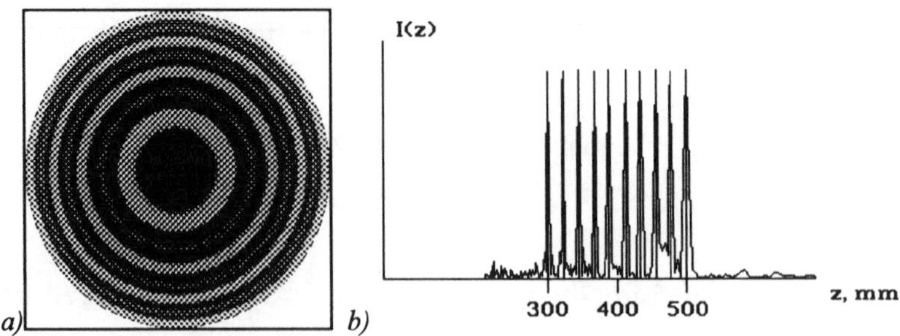

Figure 2.11 Calculation of a multifocus lens: (a) phase of a DOE; (b) axial intensity distribution

an essential advantage upon their fabrication via photolithography or e-beam lithography. In practice, such DOEs can be utilized for widening a minimum diffraction spot. This problem appears to be topical in laser printers. For example, a two-fold extension of the diffraction spot (the Airy disk) can be implemented either via a two-fold increase in the lens focal length or through the reduction of the lens aperture radius by one half. Note, however, that the former approach causes alterations in design, while the latter one reduces the efficiency of focusing by a factor of four.

We propose a 'DOE + lens' combination that offers a more effective solution to this problem.

Assume that we need to compute the DOE phase $\varphi(r)$ depending only on the radial variable, and that this DOE produces the desired intensity distribution $I_0(\rho)$ in the lens focal plane. In that case, instead of Eq. (2.6) we can write the following equation for calculating the phase $\varphi(r)$

$$I_0(\rho) = \left| \int_0^R A_0(r) \mathrm{e}^{i\varphi(r)} J_0\left(\frac{kr\rho}{f}\right) r\,\mathrm{d}r \right|^2 \tag{2.32}$$

where $A_0(r)$ is the amplitude of the illuminating beam and f is the lens focal length. Let us assume, then, that the condition $A_0(r) = 1$ is fulfilled. We will search for the $\varphi(r)$ function in a piecewise-constant form (Fig. 2.12). Then the light complex amplitude $F(\rho)$ in the lens focal plane can be represented as a sum of the amplitudes resulting from the diffraction of the plane wave by a ring-like aperture

$$F(\rho) = 2\pi \sum_{n=0}^{N-1} \mathrm{e}^{i\varphi_{n+1}} \left[r_{n+1} \frac{J_1(k\rho r_{n+1}/f)}{\rho} - r_n \frac{J_1(k\rho r_n/f)}{\rho} \right] \tag{2.33}$$

where φ_{n+1} is the value of $\varphi(r)$ at $r \in [r_n, r_{n+1}]$, $n = 0, 1, \ldots, N-1$; $r_0 = 0$, $r_N = R$, R is the DOE radius, and N is the number of partition points over the DOE radius.

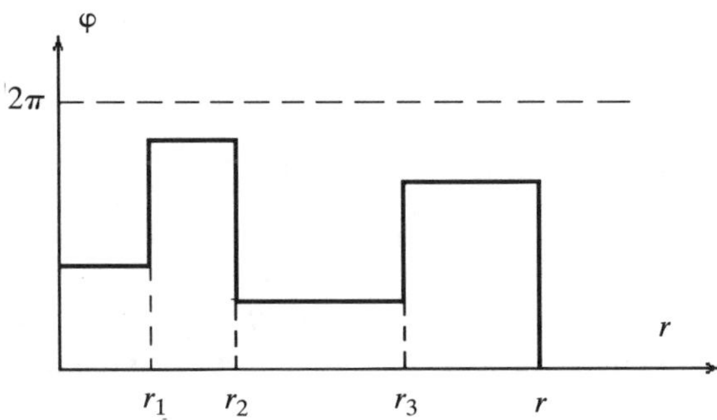

Figure 2.12 Piecewise-constant phase

On appropriate rearrangement of the terms in Eq. (2.33) we find that

$$F(\rho) = \frac{2\pi}{\rho} \sum_{n=1}^{N} C_n J_1\left(\frac{\gamma_n \rho}{a}\right) \tag{2.34}$$

where

$$\begin{cases} C_n = r_n[\mathrm{e}^{i\varphi_n} - \mathrm{e}^{i\varphi_{n+1}}] \\ C_N = R\,\mathrm{e}^{i\varphi_N} \end{cases} \tag{2.35}$$

γ_n are the zeros of the Bessel function of the first order: $J_1(\gamma_n) = 0$. It follows from a comparison of Eqs (2.33) and (2.34) that the partition points r_n should satisfy the condition

$$r_n = \frac{\gamma_n f}{ka}, \quad a \geq R \tag{2.36}$$

where a is a constant that governs the scale of partitioning of the DOE into rings.

Introducing the notation $\bar{F}(\rho) = (2\pi)^{-1}\rho F(\rho)$, instead of Eq. (2.34) we obtain

$$\bar{F}(\rho) = \sum_{n=1}^{N} C_n J_1\left(\frac{\gamma_n \rho}{a}\right) \tag{2.37}$$

Making use of the orthogonality of the Bessel function

$$\int_0^1 J_m(\gamma_p x) J_m(\gamma_q x) x \, \mathrm{d}x = \begin{cases} [J'_m(\gamma_p)]^2/2, & p = q \\ 0, & p \neq q \end{cases} \tag{2.38}$$

where $J'_m(\gamma_p)$ is the derivative of the Bessel function at point γ_p, the sums in Eq. (2.37) will be given by

$$C_n = 2[aJ_1'(\gamma_n)]^{-2} \int_0^a \bar{F}(\rho) J_1\left(\frac{\gamma_n \rho}{a}\right) \rho \, \mathrm{d}\rho \tag{2.39}$$

Relations (2.37) and (2.39) permit the following iterative procedure for solving Eq. (2.32).

1. An initial estimate $\bar{F}(\rho)$ of the complex amplitude is chosen as

$$\bar{F}(\rho) = \sqrt{[I_1(\rho)]}\, \mathrm{e}^{i\psi_0(\rho)} \tag{2.40}$$

where $I_1(\rho) = (2\pi)^{-2}\rho^2 I_0(\rho)$, $I_0(\rho)$ is the desired light intensity in the lens focal plane, and $\psi_0(\rho)$ is a random phase estimate.

2. In the kth iteration step, the function $\bar{F}_k(\rho)$ is represented as the sum of Eq. (2.37) with the coefficients $C_n^{(k)}$ that have been found using Eq. (2.39) and are replaced in the form

$$\bar{C}_n^{(k)} = B_n \mathrm{e}^{i\nu_{nk}}, \quad n = \overline{1, N} \tag{2.41}$$

where

$$B_n = 2r_n \sin\left[\frac{\varphi_n - \varphi_{n+1}}{2}\right] \tag{2.42}$$

$$\begin{cases} \varphi_n = \pi - \varphi_{n+1} + 2\nu_{nk}, & n = N-1, N-2, \ldots, 1 \\ \varphi_N = \nu_{Nk} \end{cases} \tag{2.43}$$

and $\nu_{nk} = \arg C_n^{(k)}$.

The phases φ_n are sought for use the recurrent relations in Eq. (2.43) that follow from Eq. (2.35). Actually, from Eq. (2.35) we can deduce the following sequence of equalities

$$\arg C_n = \tan^{-1}\left\{\frac{\sin(\varphi_n) - \sin(\varphi_{n+1})}{\cos(\varphi_n) - \cos(\varphi_{n+1})}\right\}$$

$$= \tan^{-1}\left\{\frac{2\sin\left(\frac{\varphi_n - \varphi_{n+1}}{2}\right)\cos\left(\frac{\varphi_n + \varphi_{n+1}}{2}\right)}{-2\sin\left(\frac{\varphi_n - \varphi_{n+1}}{2}\right)\sin\left(\frac{\varphi_n + \varphi_{n+1}}{2}\right)}\right\} \tag{2.44}$$

$$-\tan^{-1}\left\{\operatorname{ctg}\left(\frac{\varphi_n + \varphi_{n+1}}{2}\right)\right\} = -\tan^{-1}\left\{\operatorname{tg}\left(\frac{\pi}{2} - \frac{\varphi_n + \varphi_{n+1}}{2}\right)\right\}$$

$$= -\frac{\pi}{2} + \frac{\varphi_n + \varphi_{n+1}}{2}$$

from which follows the recurrent relation of Eq. (2.43).

3. A $(k+1)$th estimate $\bar{F}_{k+1}(\rho)$ is formed

$$\bar{F}_{k+1} = \sum_{n=1}^{N} \bar{C}_n^{(k)} J_1\left(\frac{\gamma_n \rho}{a}\right) \tag{2.45}$$

4. The calculated function $\bar{F}_{k+1}(\rho)$ is replaced by the function $\hat{F}_{k+1}(\rho)$ by the rule

$$\hat{F}_{k+1}(\rho) = \sqrt{[I_1(\rho)]}\,\bar{F}_{k+1}(\rho)\,|\bar{F}_{k+1}(\rho)|^{-1} \tag{2.46}$$

5. The passage to step 2 is conducted, and so forth.

The present algorithm does not converge: with increasing number of iterations, the error δ changes quasi-periodically. In practice, however, after a finite number of iterations we managed to obtain interesting results.

— *Example 2.8* —

Assume that we should compute the phase of a DOE which is able to double the diffraction spot formed by the lens in its focal plane. The parameters of calculation are: $k/f = 100\,\text{mm}^{-2}$, $R = 0.5\,\text{mm}$, and $N = 7$ is the number of

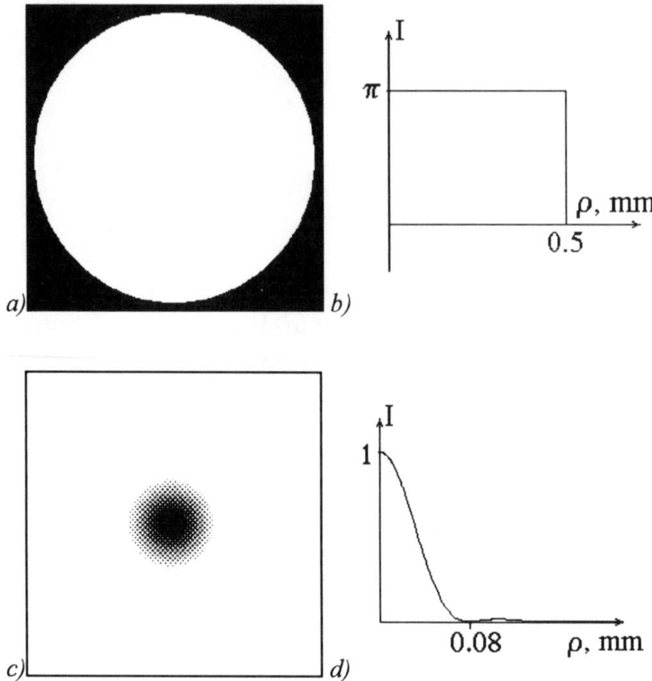

Figure 2.13 Focusing by the lens without a DOE: (a) the DOE phase and (b) its radial section; (c) the diffraction pattern and (d) its radial section

breakdowns of the DOE radius. The discreteness of the pixels along the radial variable ρ is 2 μm, with the total number of pixels equal to 256.

For calculating the coefficients through Eq. (2.39), we can use the HT of zero order in Eq. (2.7).

The desired normalized function of the intensity was chosen in the form

$$I_0(\rho) = \left[\frac{2J_1(x)}{x}\right]^2, \quad x = \frac{k\rho R}{2f} \tag{2.47}$$

which corresponds to the diffraction pattern generated by the lens with an aperture radius of $R/2$.

Figure 2.13 illustrates the result of the calculation for the lens without a DOE: the DOE phase, its radial section, the Airy disk and its radial section.

Figure 2.14 shows the result obtained after 20 iterations for the DOE with a three-level phase ($N = 3$): the phase, its radial section, diffraction pattern in the lens focal plane, and normalized intensity distribution in its horizontal section.

Note that the obtained intensity (Fig. 2.14d) differs from the desired one, Eq. (2.47), by 12% on average.

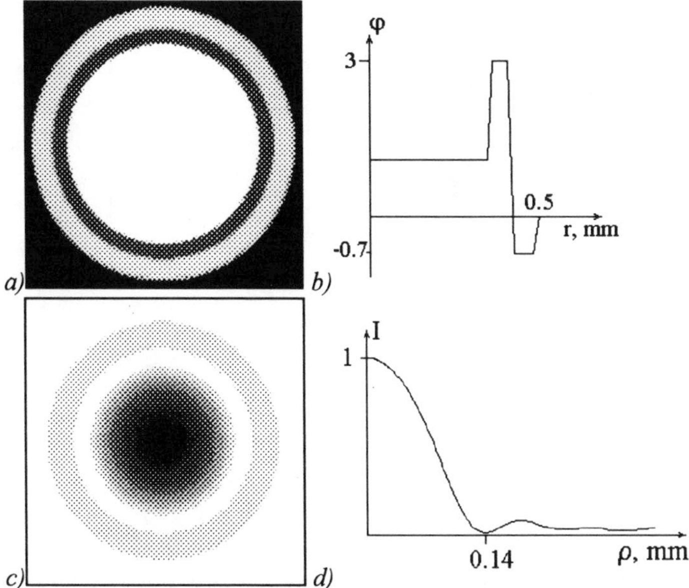

Figure 2.14 Focusing by a 'lens + DOE' combination: (a) the DOE phase and (b) its radial section; (c) the diffraction pattern and (d) its radial section

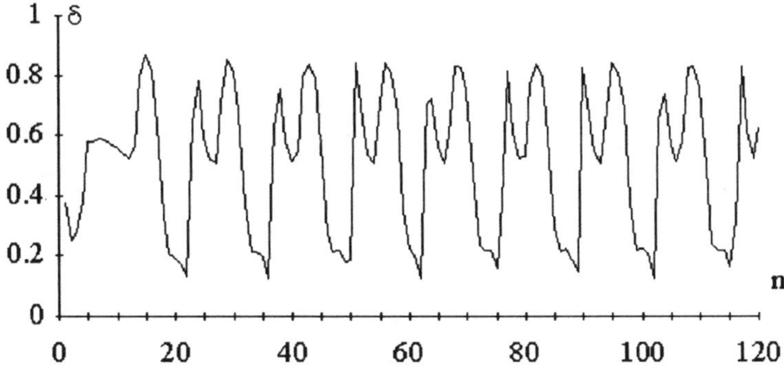

Figure 2.15 Error vs number of iterations

About 60% of the total light energy is accounted for by the ring of radius $\rho_0 = 0.14$ mm (see Fig. 2.14d). Note that a diffraction pattern analogous to that shown in Fig. 2.14c is made possible using a lens of radius $R/2$ without a DOE; however, the energy efficiency will be only 25%.

The r.m.s. error versus the number of iterations is shown for this example in Fig. 2.15. It can be seen from the figure that the plot of error has several local minima.

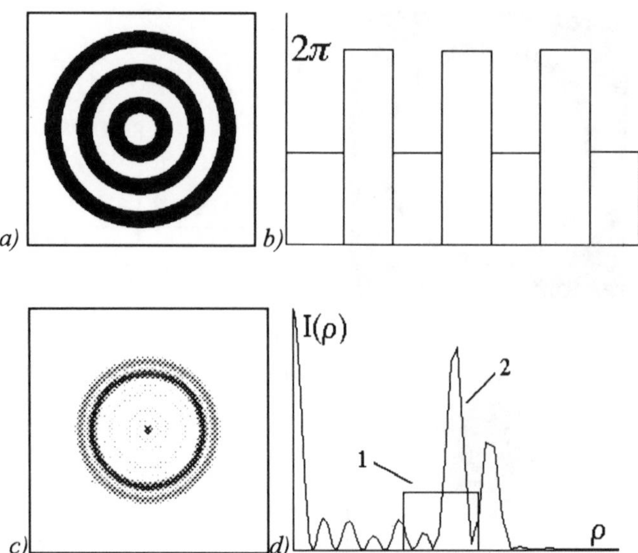

Figure 2.16 Radial DOE with two phase levels focusing into a ring: (a) the DOE phase and (b) its radial section; (c) the diffraction pattern and (d) its radial section

— *Example 2.9* ———

Let us calculate a DOE with quantized phase that can form the light into a ring [91]. The parameters of the calculation are $k/f = 100\,\text{mm}^{-2}$, $R = 0.5\,\text{mm}$ is the DOE radius, the discreteness along ρ is $2\,\mu\text{m}$, and the total number of pixels is 256.

The equation of the desired intensity is chosen in the form

$$I_0(\rho) = \text{circl}\left(\frac{\rho}{R_2}\right) - \text{circl}\left(\frac{\rho}{R_1}\right) \tag{2.48}$$

where $R_1 = 0.3\,\text{mm}$ and $R_2 = 0.5\,\text{mm}$.

Figure 2.16 depicts the result of the calculation, given that the radius is broken down into seven rings: $N = 7$. In Fig. 2.16, calculated after 12 iterations are the 2D phase distribution and its radial section (both with two quantization levels), the diffraction pattern in the lens focal plane and its radial section (curve 1 is the calculated intensity, curve 2 is the required intensity of Eq. (2.48)).

Similar results, but for $N = 11$ and $N = 15$, are shown in Figs 2.17 and 2.18, respectively. One can see that with increasing number N of the DOE radius breakdowns, the number of the phase quantization levels increases: four levels (Fig. 2.17) and five levels (Fig. 2.18). The energy efficiencies of

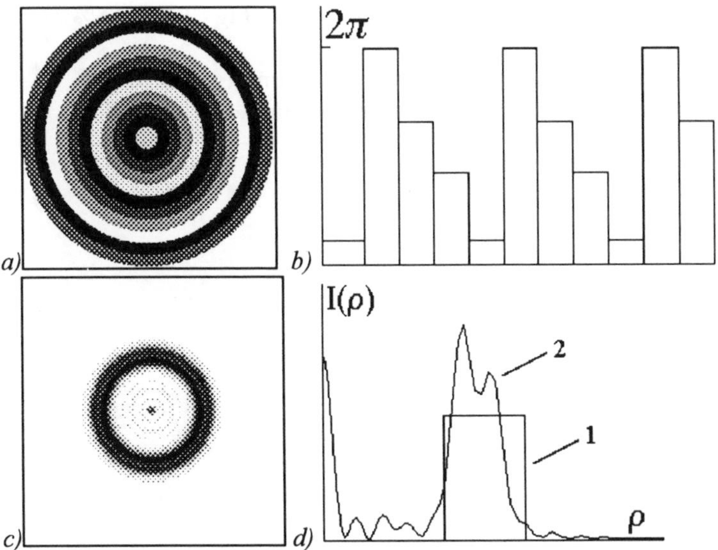

Figure 2.17 Radial DOE with four phase levels focusing into a ring: (a) the DOE phase and (b) its radial section; (c) the diffraction pattern and (d) its radial section

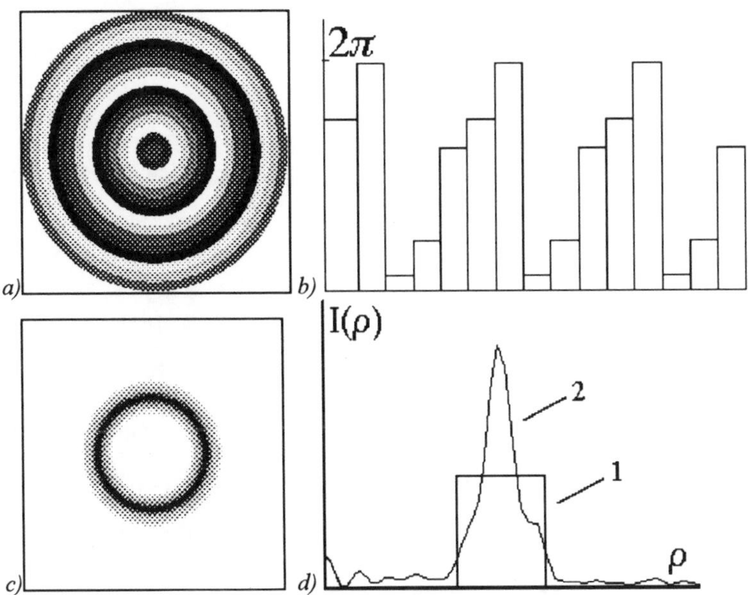

Figure 2.18 Radial DOE with five phase levels: (a) the DOE phase and (b) its radial section; (c) the diffraction pattern and (d) its radial section

focusing into a ring for this example are 60% (Fig. 2.16), 88% (Fig. 2.17), and 89% (Fig. 2.18).

It is noteworthy that, in contrast to a common practice [70], in our case the quantization of the phase is not equidistant.

3

Iterative Algorithms for Calculating Wavefront Formers

In tasks related to the non-contact checking of the shape of aspherical mirrors, one needs to synthesize phase optical elements that can form desired phase distributions in preset spatial domains [52,92]. Such optical elements are referred to as compensators and are intended to form reference wavefronts. When reflected from the surface under study, such a front is transformed into a plane wavefront (or converging spherical), provided that the surface is ideal. The quality of the surface fabrication can be judged by the deviation of the reflected wavefront from the plane one, which can be revealed by means of distortion of interferogram fringes.

The calculation of such optical elements via the methods of ray-tracing optics [52] disregard the light wave diffraction in free space. The methods of digital holography [92] enable the synthesis of a hologram which will form the desired wavefront in the first diffraction order, thus reducing the energy efficiency of the optical element.

In the subsequent text, we suggest iterative procedures by which one can calculate the formers of desired wavefronts that are not subject to the drawbacks of the above methods.

3.1 Computation of Phase Formers of Wavefronts

Let it be required to calculate the phase $\varphi(u, v)$ of an optical element which is illuminated by a light beam of amplitude $A_0(u, v)$ and forms at a distance z a light field with an arbitrary preset phase distribution $\psi_0(\xi, \eta)$. Such a task can be reduced to solving a nonlinear integral equation that, as distinct from

Eq. (1.5), takes the form [93]

$$\psi_0(\xi,\eta) = \arg\left\{ \int\int_{-\infty}^{\infty} A_0(u,v)\,\mathrm{e}^{i\varphi(u,v)} \exp\left[\frac{ik}{2z}|(u-\xi)^2+(v-\eta)^2|\right] \mathrm{d}u\,\mathrm{d}v \right\} \quad (3.1)$$

where $\arg(a+ib) = \tan^{-1}(b/a)$.

It is apparent that the procedure of iteratively solving Eq. (3.1) will differ only inessentially from the standard GS algorithm [14]: instead of replacing the calculated amplitude by the desired one, we should replace the calculated phase by the desired one. The steps of the algorithm are as follows.

1. An initial guess of the desired phase $\varphi(u,v)$ is chosen at random.
2. The function $A_0(u,v)\exp[i\varphi(u,v)]$ is Fresnel transformed (the expression in braces in Eq. (3.1)) using an algorithm of the FFT.
3. The resulting function $F_n(\xi,\eta)$ (n is the number of iterations) is replaced by $\bar{F}_n(\xi,\eta)$ according to the rule

$$\bar{F}_n(\xi,\eta) = |F_n(\xi,\eta)|\,\mathrm{e}^{i\psi_0(\xi,\eta)} \quad (3.2)$$

4. An inverse Fresnel transform of the function in Eq. (3.2) is taken, and the resulting function $W_n(u,v)$ in the DOE plane is replaced by $\bar{W}_n(u,v)$ by the conventional rule of Eq. (1.8)

$$\bar{W}_n(u,v) = \begin{cases} A_0(u,v)\,W_n(u,v)\,|W_n(u,v)|^{-1}, & (u,v)\in Q \\ 0, & (u,v)\notin Q \end{cases} \quad (3.3)$$

5. Passage to step 2.

And so forth.

The degree of convergence of the calculated phase $\psi_n(\xi,\eta)\in[0,2\pi]$ to the desired phase $\psi_0(\xi,\eta)\in[0,2\pi]$ is checked by the value of the error

$$\delta_\psi^2 = \left[\int\int_{-\infty}^{\infty} |\psi_0-\psi|^2 \mathrm{d}\xi\,\mathrm{d}\eta\right]\left[\int\int_{-\infty}^{\infty} \psi_0^2\,\mathrm{d}\xi\,\mathrm{d}\eta\right]^{-1} \quad (3.4)$$

Since the proof of convergence for this algorithm is different from that for the GS algorithm and has been derived only 'in the small', i.e. under the condition of closeness of two successive light amplitudes in the observation plane, it is presented in Appendix B. From this proof follows the inequality

$$\int\int_{-\infty}^{\infty} |A_{n+1}-A_0|^2 \mathrm{d}u\,\mathrm{d}v \le \int\int_{-\infty}^{\infty} |A_n-A_0|^2 \mathrm{d}u\,\mathrm{d}v \quad (3.5)$$

where $A_{n+1}(u,v)$ and $A_n(u,v)$ are the light amplitudes in the DOE plane calculated in the $(n+1)$th and nth iteration steps, respectively, and $A_0(u,v)$ is the preset light amplitude in the DOE plane.

From Eq. (3.5) it follows that in the course of iterations the r.m.s. deviation of the calculated amplitude from the desired one in the DOE plane decreases. We were unable to prove an analogous inequality for the phase $\psi_n(\xi, \eta)$, but all numerical simulations testify that the error of Eq. (3.4) is reduced.

— *Example 3.1* ———

To demonstrate an operation of the algorithm specified by Eqs (3.1)–(3.3), we have chosen the following parameters: the number of pixels in the DOE and in the observation planes is 128×128, the DOE radius is 5.12 mm, the pixel discreteness in these planes is 0.04 mm, $k = 10^4\,\mathrm{mm}^{-1}$, and $z_0 = 325$ mm. The amplitude of the beam illuminating the DOE was chosen to be Gaussian

$$A_0(u, v) = \exp\left[-\frac{u^2 + v^2}{w^2}\right] \tag{3.6}$$

Shown in Fig. 3.1 are the phase of a wavefront former calculated after 10 iterations, its section along the u-axis, the phase formed at the distance z_0 from the DOE and its section along the ξ-coordinate (the solid line is the desired phase; the broken line is the generated phase).

The required phase was specified by

$$\psi_0(\xi, \eta) = -\alpha(\xi^4 + \eta^4) \tag{3.7}$$

where $\alpha = 0.8$.

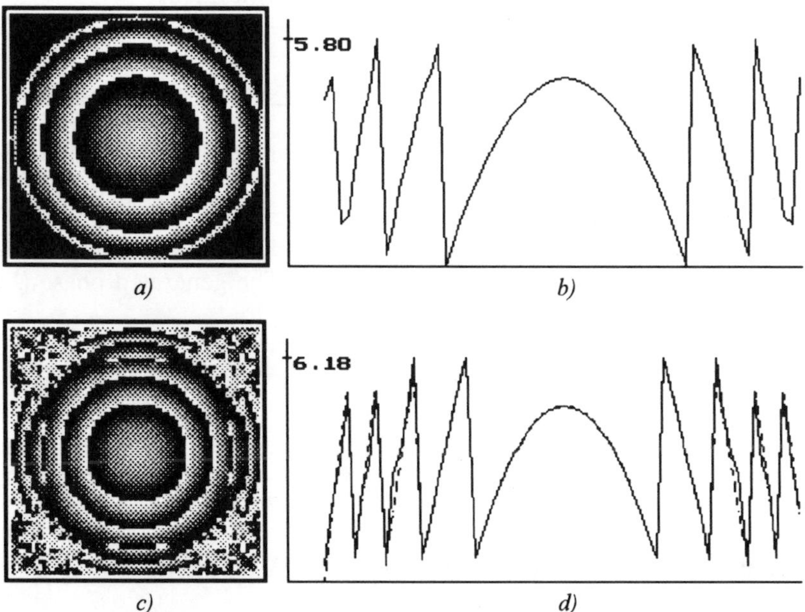

Figure 3.1 Calculation of the phase former of a wavefront: (a) the DOE phase and (b) its horizontal section; (c) the generated phase and (d) its horizontal section

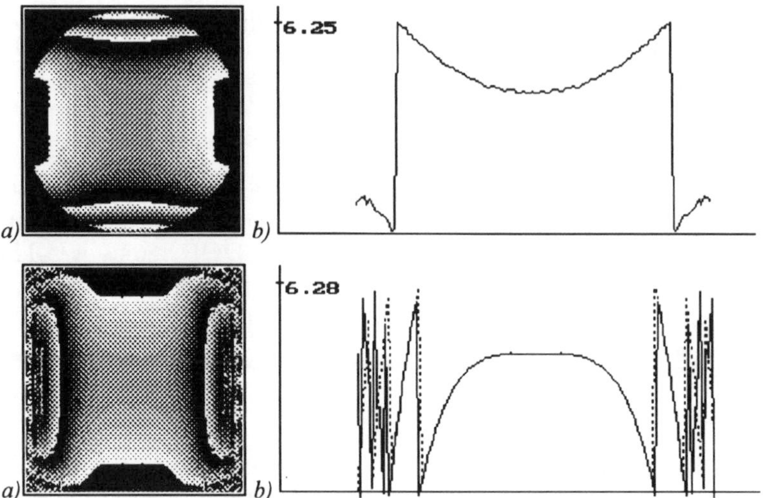

Figure 3.2 Calculation of the phase former of a wavefront: (a) the DOE phase and (b) its horizontal section; (c) the generated phase; (d) its horizontal section

The deviation of the calculated phase $\psi_n(\xi, \eta)$ from the required one in Eq. (3.7) that has been calculated using Eq. (3.4) is 12%.

— *Example 3.2* —

This method can be applied to the calculation of a DOE able to form a light field with different curvatures along the u- and v-axes. For example, Fig. 3.2 depicts the DOE phase, its section along the u-axis, the phase produced at the distance z_0, and its section along the ξ-axis (the solid line is the desired phase, the broken line is the generated phase). The generated phase (Fig. 3.2c) differs from the required one given by

$$\psi_0(\xi, \eta) = -\beta(\xi^2 - \eta^4) \tag{3.8}$$

where $\beta = 0.46$ by 20% on average.

3.2 Amplitude Transparencies for Forming Wavefronts

This section deals with the problem of calculating an amplitude transparency (AT) that transforms a plane coherent wave into a preset phase distribution at some distance along the optical axis. In a particular case, such an AT can be treated as an amplitude lens that transforms an incident plane wavefront into a converging wavefront, thus providing the concentration of light energy on an optical axis at some distance from the AT. The problem of an AT-based focusing of laser light is dealt with in [94].

In calculating a DOE, such parameters as the amplitudes in the planes of an optical element and the observation phase are taken to be known, whereas the DOE phase is sought. In what follows, we consider an 'inverse' task: the light field phase in the AT plane and in the observation plane is taken to be known, and the AT transmission is sought.

This problem turns out to be formally equivalent to solving the integral equation

$$\psi_0(\xi, \eta) = \arg\left\{ \int\int_{-\infty}^{\infty} A(u, v) e^{i\varphi_0(u,v)} \exp\left[\frac{ik}{2z}|(u-\xi)^2 + (v-\eta)^2|\right] du\, dv \right\} \quad (3.9)$$

where $\psi_0(\xi, \eta)$ and $\varphi_0(u, v)$ are the pregiven phases in the observation plane located at a distance z from the AT and in the transparency plane, respectively, and $A(u, v)$ is the desired function of the transparency amplitude transmission.

The difference between Eqs (3.9) and (3.1) implies that in the integrand of Eq. (3.1) the phase is considered to be unknown, whereas in Eq. (3.9) it is the amplitude $A(u, v)$ that is unknown. Therefore, the algorithm for solving Eq. (3.9) will follow the same steps as the algorithm for solving Eq. (3.1), but in place of the replacement of Eq. (3.3), one should use the following replacement [95]

$$\bar{W}_n(u, v) = \begin{cases} |W_n(u, v)| e^{i\varphi_0(u,v)}, & (u, v) \in Q \\ 0, & (u, v) \notin Q \end{cases} \quad (3.10)$$

where Q is the shape of the AT aperture.

If the transparency is illuminated by the plane wave, $\varphi_0(u, v)$ will be constant. In the case of such an iterative algorithm for solving Eq. (3.9) we can prove the convergence on the average. Note that for any n (n is the number of iterations) there holds the inequality

$$\int\int_{-\infty}^{\infty} |A_{n+1}(u, v) - A_n(u, v)|^2 du\, dv \leq \int\int_{-\infty}^{\infty} |A_n(u, v) - A_{n-1}(u, v)|^2 du\, dv \quad (3.11)$$

which shows that in the course of iterations the r.m.s. deviation for two successive amplitudes in the AT plane will not increase.

In a similar way, one can show that for the field amplitudes in the observation plane the following inequality is valid

$$\begin{aligned}&\iint_{-\infty}^{\infty} |B_{n+1}(\xi,\eta) - B_n(\xi,\eta)|^2 \mathrm{d}\xi\mathrm{d}\eta \\ &\le \iint_{-\infty}^{\infty} |B_n(\xi,\eta) - B_{n-1}(\xi,\eta)|^2 \mathrm{d}\xi\mathrm{d}\eta\end{aligned} \tag{3.12}$$

where B_{n+1}, B_n and B_{n-1} are the light field amplitudes in the observation plane in the corresponding iteration steps. The proof of inequality (3.12) is given in Appendix C.

We failed to prove that in the course of iterations the r.m.s. error of the phase in Eq. (3.4) diminishes, but using particular numerical examples we managed to show that in the points (ξ, η) where the amplitude $B_n(\xi, \eta)$ is found to be close to zero, the phase $\psi_n(\xi, \eta)$ can be essentially different from the desired phase $\psi_0(\xi, \eta)$. By this is meant that the amplitude convergence resulting from Eq. (3.12) leads to the phase convergence, while the function to which the $\psi_n(\xi, \eta)$ phase converges can be essentially different from the desired phase in the points of the field where the light energy is close to zero.

— *Example 3.3* ———

Figure 3.3 depicts the transmission function of the AT calculated using the iterative algorithm within 800 iterations, the phase which is formed by the transparency at distance $z_0 = 20\,\mathrm{mm}$ (solid line) and is different from the desired phase (broken line) in the points of small intensity, and the intensity distribution that is formed by this transparency at a distance $z = 900\,\mathrm{mm}$. Parameters of the calculation are $k = 10^4\,\mathrm{mm}^{-1}$, the pixel discreteness in the AT plane is 0.01 mm, and the number of pixels is 128.

The difference between the synthesized phase (Fig. 3.3b) and the desired one has a specific character: in the points where the intensity distribution is essentially non-zero (the intensity distribution in the observation plane is just the same as in the plane of the AT itself, see Fig. 3.3a) the agreement between the calculated and the desired phases is quite satisfactory, but in the points where the field intensity is near to zero an essential difference between these phases occurs.

The desired phase function was chosen in the form

$$\psi_0(\xi) = -\alpha\xi^2 \tag{3.13}$$

where $\alpha = 2\,\mathrm{mm}^{-2}$.

Note that we can understand, at least qualitatively, the relation between the intensity distribution on the transparency (Fig. 3.3a) and the field

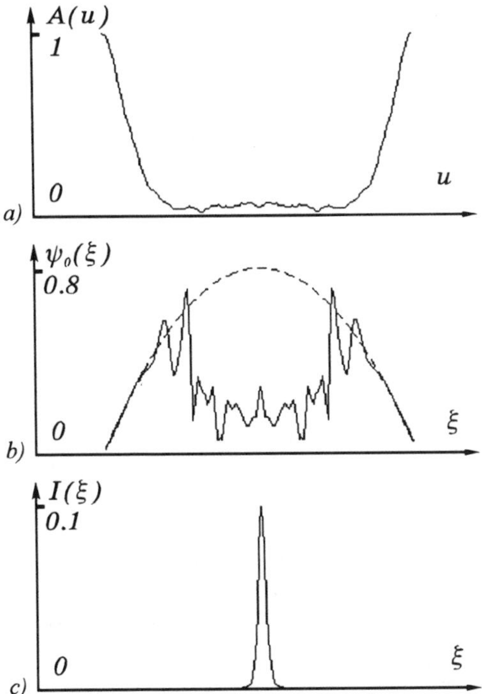

Figure 3.3 Calculation of an amplitude transparency: (a) the AT amplitude; (b) the generated phase (solid line) and the desired phase (broken line); (c) the far field intensity

amplitude at a large distance from the transparency (Fig. 3.3c) from the following familiar integral

$$\int_{-a}^{a} [a^2 - x^2]^{-1/2} e^{ix\xi} dx = \pi J_0(a\xi) \tag{3.14}$$

where $J_0(x)$ is the zero-order Bessel function.

Figure 3.4 depicts the r.m.s. deviation of Eq. (3.12) versus the number of iterations for the field amplitude in the plane of formation of the desired wavefront. It is seen that the error does not rise with increasing number of iterations, which is in agreement with the proof presented in Appendix C.

— *Example 3.4*

The iterative method can be applied with more apparent success to the calculation of an AT forming diverging wavefronts. Figure 3.5 depicts the AT transmission function obtained after 100 iterations (with the same

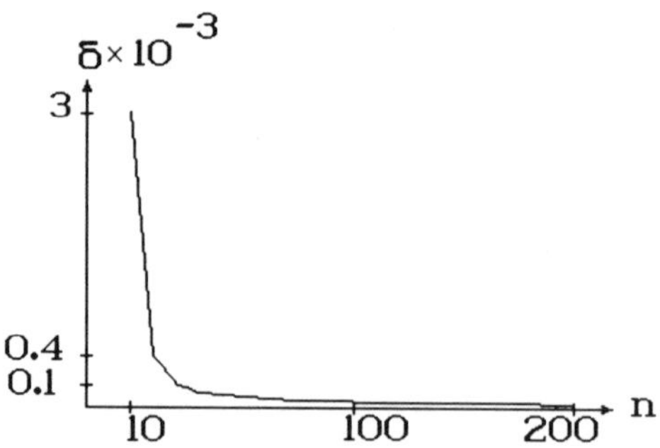

Figure 3.4 Error vs number of iterations

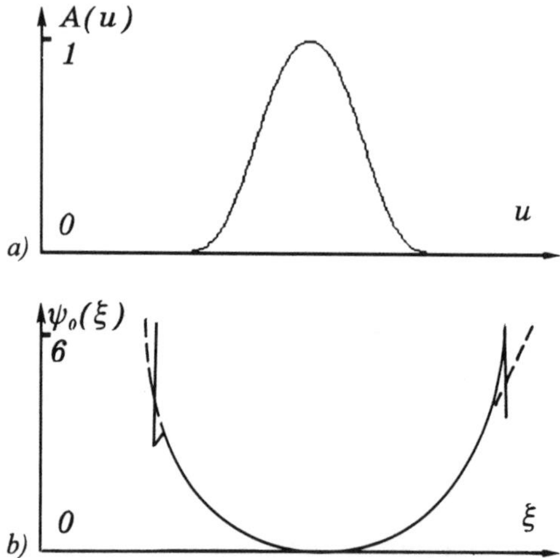

Figure 3.5 Calculation of an AT: (a) the AT amplitude; (b) the generated phase

parameters of calculation) and the phase (solid line) which is formed by the transparency at a distance of z_0 and is almost identical to the required phase (broken line)

$$\psi_0(\xi) = \beta\xi^4 \tag{3.15}$$

where $\beta = 17\,\text{mm}^{-4}$.

Let us now consider another particular modernization of the iterative GS algorithm [14] that is better suited for calculating amplitude lenses. By the amplitude lens we mean an AT that ensures the concentration of light energy in a small region of the observation plane located at some distance.

In that case, instead of Eq. (3.9) we need to solve iteratively the equation

$$I_0(\xi, \eta) = \left| \frac{k}{z} \int \int_{-\infty}^{\infty} A(u, v) \exp\left[\frac{ik}{2z} |(u-\xi)^2 + (v-\eta)^2| \right] \mathrm{d}u\,\mathrm{d}v \right|^2 \quad (3.16)$$

where $I_0(\xi, \eta)$ is the desired intensity distribution in the observation plane and $A(u, v)$ is the unknown amplitude transmittance of the lens. To ensure the concentration of light energy into a small region of the observation plane, we propose choosing the intensity in the form

$$I_0(\xi, \eta) = p\,\mathrm{rect}\left(\frac{\xi}{T_1}\right) + q\,\mathrm{rect}\left(\frac{\xi}{T_2}\right) \quad (3.17)$$

where $T_1 = \lambda z/2a$ is the radius of a small diffraction spot in the observation plane, $T_2 = a$, $2a$ is the aperture size of the AT, and $p \gg q$. In Eq. (3.17) the first term approximates the central peak of intensity in the plane of focusing, and the second term approximates an inevitable background.

Equation (3.16) can be solved via iterations in a similar way to the procedure of solving Eq. (1.5) using the GS algorithm, but in this case instead of the replacement in Eq. (1.8) we should use the replacement given by

$$\bar{W}_n(u, v) = \begin{cases} |W_n(u, v)|, & (u, v) \in Q \\ 0, & (u, v) \notin Q \end{cases} \quad (3.18)$$

where $W_n(u, v)$ is the light complex amplitude in the AT plane obtained in the nth iteration step and Q is the aperture shape of the AT.

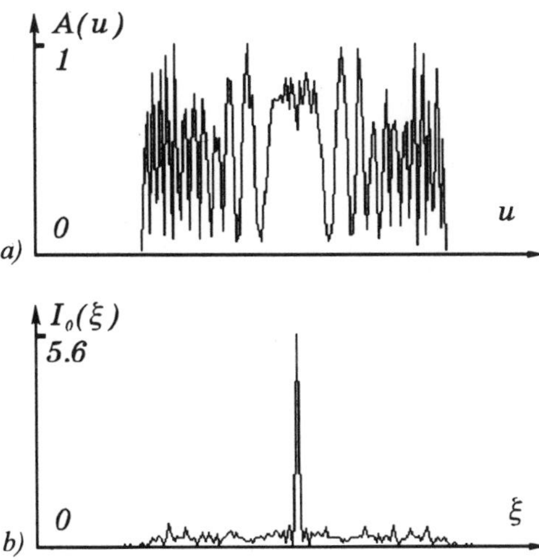

Figure 3.6 Calculation of an amplitude lens: (a) the AT amplitude; (b) the generated intensity

— *Example 3.5* —

Figure 3.6 (see previous page) shows the transparency amplitude transmittance calculated after 20 iterations and the intensity distribution formed by this transparency at a distance $z_0 = 20$ mm. By fitting the constants p and q we can attain in Eq. (3.17) a maximum possible energy density in the central peak for the given parameters. In Fig. 3.6 the intensity of the central peak is 5.6 times larger than the maximum unity intensity of the field near the transparency and is 50 times larger than an average value of the background intensity in the plane of focusing.

The efficiency of focusing in this case amounts to 5%. By the efficiency we mean the portion of light energy of the plane beam illuminating the transparency that is spent on the formation of the central peak in the plane of focusing. The diffraction efficiency is the ratio of the light energy going to the generation of the central peak to the entire energy having passed through the AT. In Fig. 3.6 the diffraction efficiency is 17%.

4

Calculation of Phase Formers of Light Modes

4.1 Bessel Mode Formers

References [54,96,97] report studies of light beams which can be referred to as Bessel modes of zero order since the amplitude of the beams is proportional to Bessel functions of the first kind and of zero order. It is mentioned in [98] that if the amplitude of a beam is proportional to Bessel functions of nth order, the beam propagates in free space without diffraction. Reference [55] presents analysis, simulation and experimental results on the production of nondiffracting beams involving nth order Bessel functions. The approach of [55] is different from the algorithm presented here, and deals with the filtering of the phase and amplitude of an incident wave. From [99], light Bessel functions are known to describe the complex amplitude of light within a circular fibre core as well as the light amplitude at the output mirror of a plane-parallel cavity with circular mirrors.

Reference [100] proposes an algorithm for calculating phase optical elements forming quasimodes of free space, i.e. light beams that effectively represent a superposition of a small number of Bessel functions of the same order but with different arguments. It is shown in [50] that a conic axicon [87], when taken separately or in combination with a phase rotor filter [64], is an example of an optical phase element forming quasimodes of zero or first order, respectively.

In the present chapter we have developed an algorithm [30] for iteratively calculating phase optical elements capable of forming nondiffracting light beams with an arbitrary Bessel mode composition.

Reference [101] reports the derivation of the general expression for the field complex amplitude which remains unchanged with distance to an accuracy of a phase factor depending on the axial coordinates z. This expression is given by

$$U(r,\varphi,z)=\frac{1}{2\pi}\exp(ikz\sigma)\int_0^{2\pi}F(\theta)\exp[-ikr\rho_0\cos(\theta-\varphi)]\,\mathrm{d}\theta \tag{4.1}$$

where $\sigma=\sqrt{(1-\rho_0^2)}$, $U(r,\varphi,z)$ is the complex amplitude of the nondiffracting beam, $F(\theta)$ is an arbitrary function of the polar angle, k is the wavenumber, (r,φ) are the polar coordinates, and z is the coordinate coaxial to the propagation of the nondiffracting beam. Equation (4.1) shows that Bessel modes are a superposition of plane waves with the fixed angle of inclination ν to the optical axis: $\nu=\sin^{-1}(\rho_0)$.

If the function $F(\theta)$ is periodically dependent on the angle θ, it can be expanded into a Fourier series

$$F(\theta)=\sum_{n=-\infty}^{\infty}C_n\exp(-in\theta) \tag{4.2}$$

Substitution of Eq. (4.2) into Eq. (4.1) gives

$$U(r,\varphi,z)=\exp(ikz\sigma)\sum_{n=-\infty}^{\infty}(-i)^nC_nJ_n(r\beta)\exp(-in\varphi) \tag{4.3}$$

where $\beta=k\rho_0$. In deriving Eq. (4.3) we made use of the integral form of the Bessel function

$$J_n(x)=\frac{i^n}{2\pi}\int_0^{2\pi}\exp(-int)\exp(-ix\cos t)\,\mathrm{d}t \tag{4.4}$$

To form the light field of Eq. (4.3), one needs to form in the plane $z=0$ the field

$$U_0(r,\varphi)=\sum_{n=-N}^{N}\tilde{C}_nJ_n(r\beta)\exp(-in\varphi) \tag{4.5}$$

where $2N+1$ is the number of terms in the sum, $\tilde{C}_n=(-i)^nC_n$.

We can check that the light field in Eq. (4.5) is conserved with propagation along the optical axis. For this purpose, the Fresnel integral in polar coordinates needs to be computed

$$U(\rho,\psi,z)=\frac{-ik}{2\pi z}\mathrm{e}^{ikz}\int_0^{\infty}\int_0^{2\pi}U_0(r,\varphi)\,\mathrm{e}^{i(k/2z)(r^2+\rho^2)}\mathrm{e}^{-i(k/z)r\rho\cos(\psi-\varphi)}r\,\mathrm{d}r\,\mathrm{d}\varphi \tag{4.6}$$

where $U(\rho,\psi,z)$ is the complex amplitude in cylindrical coordinates. Substituting Eq. (4.5) into Eq. (4.6), we find

$$\begin{aligned}U(\rho,\psi,z)=&\frac{-ik}{z}\,\mathrm{e}^{ikz}\,\mathrm{e}^{i(k/2z)\rho^2}\sum_{n=-N}^{N}(-i)^n\tilde{C}_n\mathrm{e}^{-in\psi}\\&\times\int_0^{\infty}J_n\left(\frac{k}{z}r\rho\right)J_n(r\beta)\,\mathrm{e}^{i(k/2z)r^2}r\,\mathrm{d}r\end{aligned} \tag{4.7}$$

Making use of a tabulated integral ([102], p. 227)

$$\int_0^\infty J_n(bx)J_n(cx)\exp(iax^2)x\,\mathrm{d}x = \frac{i^{n+1}}{2a}J_n\left(\frac{bc}{2a}\right)\exp\left(-i\frac{b^2+c^2}{4a}\right) \quad (4.8)$$

instead of Eq. (4.7) we obtain

$$U(\rho,\psi,z) = \exp\left[iz\left(k-\frac{\beta^2}{2k}\right)\right]\sum_{n=-N}^{N}\tilde{C}_nJ_n(r\beta)\exp(-in\psi) \quad (4.9)$$

which differs from Eq. (4.5) only by a z-dependent factor before the sum sign. Note that Eq. (4.9) is equivalent to Eq. (4.3) given that the condition of paraxiality

$$\sqrt{(1-\rho_0^2)} = 1-\frac{\rho_0^2}{2}$$

is satisfied. The final statement of the calculation task is as follows: one needs to calculate a phase optical element with the phase function $\exp[i\psi(r,\Theta)]$ which transforms the plane light wave into a nondiffracting beam with a pregiven Bessel mode composition. That is, the following equality is to be obeyed

$$\exp[i\psi(r,\varphi)] = \sum_{n=-N}^{N}C_nJ_n\left(\gamma_{n0}\frac{r}{R}\right)\exp(-in\varphi) \quad (4.10)$$

where γ_{n_0} are the roots of a Bessel function, $J_n(\gamma_{n_0})=0$, closest to the pregiven value βR; $\gamma_{n_0}\approx\beta R$, R is the optical element radius, and β is a parameter connected with the beam effective diameter and with a distance z_0 up to which the beam propagates without a distortion, which arises from the finite aperture [96]

$$z_0 = R\sqrt{\left[\left(\frac{k}{\beta}\right)^2-1\right]} \quad (4.11)$$

Absolute values of the coefficients $|C_n| = B_n$ are chosen arbitrarily, with the coefficients' arguments, $\nu_n = \arg C_n$, taken as free parameters.

We propose that Eq. (4.10) should be solved using an iterative technique. An initial estimate $\Psi_0(r,\varphi)$ of a desired phase is chosen. Using the function $U_0(r,\varphi) = \exp[i\Psi(r,\varphi)]$ and the relationship

$$C_n = A_n\int_0^R\int_0^{2\pi}U_0(r,\varphi)J_n\left(\gamma_{n0}\frac{r}{R}\right)\exp(in\varphi)r\,\mathrm{d}r\,\mathrm{d}\varphi \quad (4.12)$$

where $A_n = \{\pi[RJ_n'(\gamma_{n_0})]^2\}^{-1}$, $n=-N,N$, and $J_n'(x)$ are the derivatives of the Bessel function, we derive the coefficients of the sum in Eq. (4.10). Note that Eq. (4.12) follows from the orthogonality of the functions $\exp(in\theta)$

$$\int_0^{2\pi}\exp[i\theta(n-m)]\,\mathrm{d}\theta = 2\pi\delta_{nm} \quad (4.13)$$

(where δ_{nm} is the Kronecker delta) and from the well-known normalizing integral for Bessel functions

$$\int_0^R J_n^2\left(\gamma_{n0}\frac{r}{R}\right)r\,\mathrm{d}r = \frac{1}{2}[RJ_n'(\gamma_{n0})]^2 \tag{4.14}$$

Next, the calculated coefficients C_n are replaced by the $\bar{C}_n$ coefficients using the rule

$$\bar{C}_n = B_n \exp(i \arg C_n) \tag{4.15}$$

where B_n are the arbitrary positive numbers characterizing the Bessel mode composition of the nondiffracting beam.

Based on the coefficients $\bar{C}_n$, the sum in Eq. (4.10) is computed whose argument is a new estimate of the phase $\Psi(r, \varphi)$

$$\Psi_1(r,\varphi) = \arg\left\{\sum_{n=-N}^{N} \bar{C}_n J_n\left(\gamma_{n0}\frac{r}{R}\right)\exp(-in\varphi)\right\} \tag{4.16}$$

and so forth.

The convergence of the iterative algorithm can be checked by the deviation δ_n of the absolute values $|C_n|$ of calculated coefficients from the pregiven numbers B_n

$$\delta_n^2 = \frac{1}{(2N+1)^2}\sum_{n=-N}^{N}[|C_n| - B_n]^2 \tag{4.17}$$

To calculate the coefficients C_n using Eq. (4.12), the fast Fourier transform is employed. First, $L > 2N+1$ one-dimensional Fourier transforms are computed

$$I_n(r) = \int_0^{2\pi} U_0(r,\varphi)\mathrm{e}^{in\varphi}\mathrm{d}\varphi \approx \frac{2\pi}{M}\sum_{p=1}^{M} U_p(r)\mathrm{e}^{2\pi i(pn/M)} \tag{4.18}$$

where $M \geq 2N+1$, $U_p(r) = U_0(r, \varphi_p)$, $\varphi_p = 2\pi(p/M)$, M is the number of breakdowns of the azimuth angle $\varphi \in [0, 2\pi]$, and L is the number of breakdowns of the variable $r \in [0, R]$.

Next, $2N+1$ Hankel transforms are taken

$$C_n = A_0 \int_0^R I_n(r) J_n\left(\gamma_{n0}\frac{r}{R}\right) r\,\mathrm{d}r \tag{4.19}$$

Integrals of Eq. (4.19) can also be computed using a Fourier transform [28,81].

It can be shown that this iterative algorithm for calculating phase formers of nondiffracting beams converges in the average, which means that for any number p the following equality is obeyed

$$\sum_{n=-\infty}^{\infty} |C_n^{(p+1)}| B_n \geq \sum_{n=-\infty}^{\infty} |C_n^{(p)}| B_n \tag{4.20}$$

where $C_n^{(p+1)}$ and $C_n^{(p)}$ are the coefficients calculated in the $(p+1)$th and pth iteration steps, respectively. The proof of the inequality (4.20) is similar to that reported in [103], where instead of the sum in Eq. (4.10) an expansion in terms of orthogonal Hermite polynomials was employed. The proof of inequality (4.20) is given in Appendix D [30].

— *Example 4.1*

The parameters of numerical simulation are as follows. The number of gradations along the radial variable is $L = 64$, the number of gradations along the azimuth angle is $M = 128$, $R = 1$ mm is the radius of an optical element, $\beta = 52.6\,\text{mm}^{-1}$, and $k = 2\pi/\lambda = 10^4\,\text{mm}^{-1}$. The number of terms in the sum of Eq. (4.10) equals $2N + 1 = 11$.

Figure 4.1a depicts the optical element phase in which calculation the following values of the coefficients have been chosen

$$B_n = \begin{cases} 1, & n = 0 \\ 0, & \text{else} \end{cases} \tag{4.21}$$

This means that an optical element with the phase shown in Fig. 4.1a forms the light beam described by the Bessel function of zero order. This is a conventional binary axicon. Figure 4.1b shows the calculated transverse intensity distributions (grey-level, negative) formed by this optical element for different z: 80 mm, 100 mm, 120 mm. The beam effective radius is about

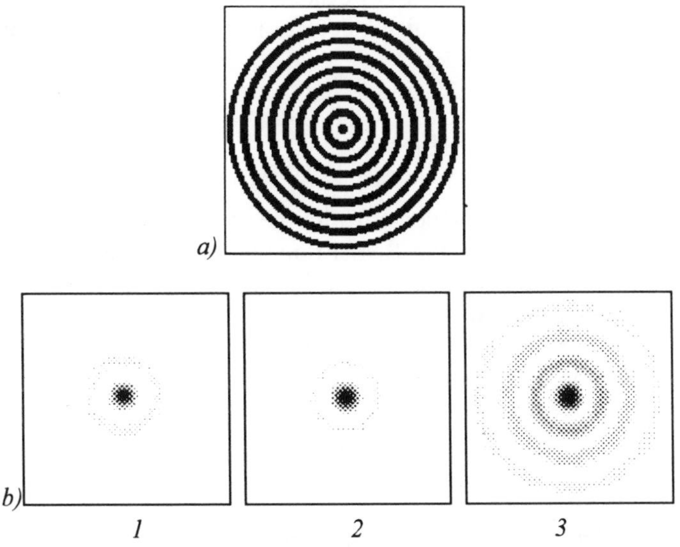

Figure 4.1 (a) Calculated phase of a binary axicon forming a beam with transverse intensity distribution as shown in (b): (1) $z = 80$ mm; (2) $z = 100$ mm; (3) $z = 120$ mm

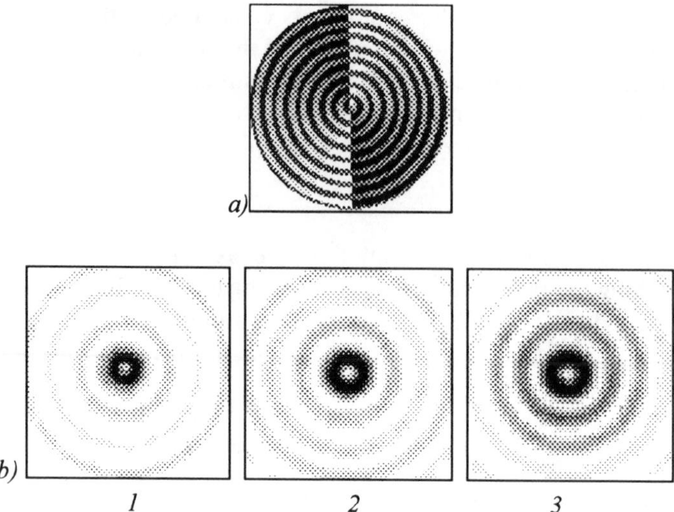

Figure 4.2 (a) Calculated phase of an optical element forming a 'pipe' with transverse intensity distribution as shown in (b): (1) $z = 80$ mm; (2) $z = 100$ mm; (3) $z = 120$ mm

120 μm. It is seen from Fig. 4.1b that the light beam propagates with almost no change in its structure.

Figure 4.2a depicts the phase of optical elements (grey-level, 16 gradations within the range $[0, 2\pi]$), in which calculation the following absolute values of coefficients have been used

$$B_n = \begin{cases} 1, & n = 1 \\ 0, & \text{else} \end{cases} \tag{4.22}$$

This means that an optical element with the phase shown in Fig. 4.2a forms the light beam described by the Bessel function of the first order. It is seen from Fig. 4.2b that the light beam in the form of a 'pipe' possesses zero axial intensity and propagates with almost no change in its structure.

Figure 4.3a illustrates the phase for which calculation instead of Eqs (4.21) and (4.22) the following numbers were used

$$B_n = \begin{cases} 1, & n = -1, 0, 1 \\ 0, & \text{else} \end{cases} \tag{4.23}$$

Figure 4.3b depicts transverse intensity distributions (negative) formed by the optical element with such a phase for the same values of the z-coordinate as is the case of Fig. 4.1b. It is seen from Fig. 4.3b that the light beam in the form of an 'oblate cylinder' axially propagates with almost no change.

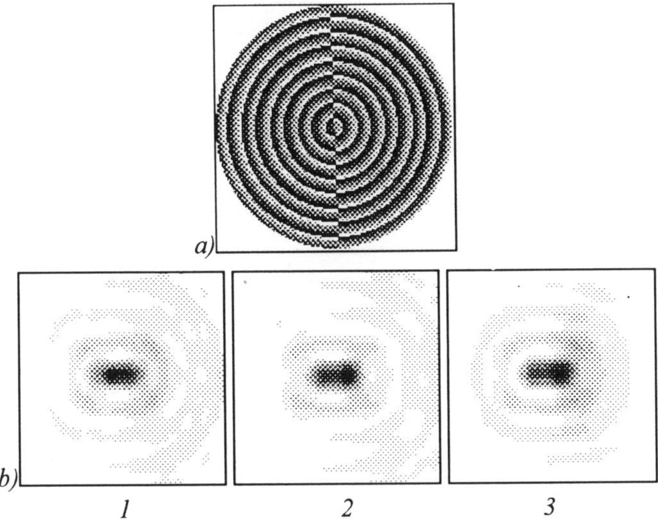

Figure 4.3 (a) The phase of the DOE forming an 'oblate cylinder' with transverse intensity distribution as shown in (b): (1) $z = 80$ mm; (2) $z = 100$ mm; (3) $z = 120$ mm

Figure 4.4a illustrates the phase in which calculation the following absolute values of coefficients have been used

$$B_n = \begin{cases} 1, & n = -1, 1 \\ 0, & \text{else} \end{cases} \tag{4.24}$$

The intensity distribution of the beam is shown in Fig. 4.4b for the same z as for the foregoing cases. The light field in the form of two parallel light beams is seen from Fig. 4.4b to be almost unchangeable with distance. An optical element with the phase shown in Fig. 4.4a can be called a 'bi-axicon', as it forms two nondiffracting beams.

Figure 4.5a illustrates the phase in which calculation the following absolute values of coefficients have been used

$$B_n = \begin{cases} 1, & n = -2, -1, 0, 1, 2 \\ 0, & \text{else} \end{cases} \tag{4.25}$$

The light field in the form of two parallel light beams is seen from Fig. 4.5b to be almost unchangeable with distance. In this case the distance between the light beams is more than the previous one.

The iterative algorithm developed in this section makes it possible to calculate phase optical elements which then can be fabricated using laser- or e-beam lithography technology. The elements can form nondiffracting light beams with a pregiven Bessel composition, i.e. with a complex form of transverse intensity distribution, namely a circle, an ellipse, two circles,

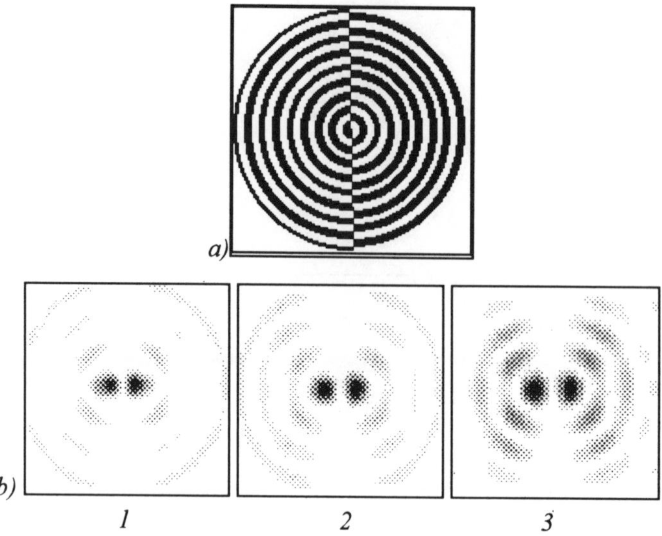

Figure 4.4 (a) Calculated phase of a 'bi-axicon' forming a beam with transverse intensity distribution as shown in (b): (1) $z = 80$ mm; (2) $z = 100$ mm; (3) $z = 120$ mm

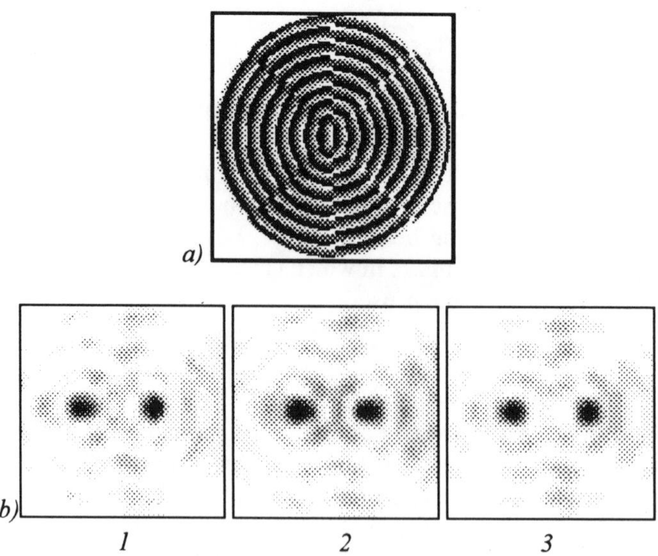

Figure 4.5 (a) Calculated phase of the DOE forming a beam with transverse intensity distribution as shown in (b): (1) $z = 80$ mm; (2) $z = 100$ mm; (3) $z = 120$ mm

and the like. Such optical elements can be treated as generalized axicons and applied to the usual tasks for the conic axicon applications.

4.2 Formers of Bessel Quasimodes

As noted above, Refs [96,97] treated the nondiffracting beam with an amplitude proportional to the zero-order Bessel function. To produce such a beam (of the zero Bessel mode), one needs either to synthesize an optical element with an amplitude-phase transmittance equal to the $J_0(x)$ function, or to illuminate a narrow, ring-like slot $\beta f/k$ in radius by the plane beam and to locate a spherical lens with focal length f at a distance f from the slot [97].

Both these approaches have obvious practical disadvantages. To gain better insight into how one can calculate a purely phase optical element capable of 'effectively' forming the zero Bessel mode, we introduce the term 'quasimode' [100].

The light field being a superposition of a small number of zero Bessel modes with different widths will be referred to as the quasimode

$$\hat{F}_0(r) = \sum_{n=K}^{L} C_n J_0\left(\gamma_n \frac{r}{R}\right) \tag{4.26}$$

where γ_n are the Bessel function roots, $J_0(\gamma_n) = 0$ and R is the DOE radius. The central root with number $n_0 = (K + L)/2$ is chosen from the condition $\gamma_n = \beta R$, where β is a constant in the argument of $J_0(\beta r)$ that characterizes the radius r_0 of the zero Bessel quasimode

$$r_0 = \frac{\gamma_1}{\beta} \tag{4.27}$$

γ_1 is the first zero of the zero-order Bessel function: $\gamma_1 = 2.4$.

As the beam of Eq. (4.26) propagates in free space, the appearance of the entire beam and its effective radius will alter, because in propagating each mode, each term in Eq. (4.26) has its own phase speed. The degree of diffraction divergence of the quasimode beam is proportional to the number of terms in Eq. (4.26).

If we resort to the property of orthogonality of the Bessel functions of the same order

$$\int_0^1 J_n(x\gamma_p) J_n(x\gamma_q) x \, dx = \frac{1}{2} [J'_n(\gamma_p)]^2 \delta_{pq} \tag{4.28}$$

we can show that the coefficient C_n in Eq. (4.26) may be found from the relation

$$C_n = 2[RJ'_0(\gamma_n)]^{-2} \int_0^R \hat{F}_0(r) J_0\left(\gamma_n \frac{r}{R}\right) r \, dr \tag{4.29}$$

Then the algorithm for iteratively searching for the phase of the optical element producing zero-order quasimodes, i.e. weakly diverging beams being a superposition of a small number of zero-order Bessel modes similar in their effective radii, is as follows.

In the kth step, an estimate of the optical element transmittance is formed

$$\hat{F}_k(r) = A_0(r)\mathrm{e}^{i\varphi_k(r)} \tag{4.30}$$

where $A_0(r)$ is the amplitude of the illuminating light beam and $\varphi_k(r)$ is the DOE phase derived in the kth iteration step.

Next, the coefficients $C_n^{(k)}$ are deduced using Eq. (4.29) and are replaced by $\bar{C}_n^{(k)}$ by the rule

$$\bar{C}_n^{(k)} = \begin{cases} B_n \mathrm{e}^{i \arg C_n^{(k)}}, & n \in [K, L] \\ 0, & n \notin [K, L] \end{cases} \tag{4.31}$$

The subsequent estimate $\hat{F}_{k+1}(r)$ is found from summation of Eqs (4.26) and (4.31). This done, the argument of the calculated complex amplitude remains unchangeable, with the module to be replaced by a pregiven function, as is the case in Eq. (4.30), and so forth.

The algorithm that involves Eqs (4.26) and (4.29)–(4.31) is similar to that for calculating Bessel mode formers described in section 4.1, but permits a simpler software implementation. The proof of the algorithm convergence is analogous to that given in Appendix E.

— *Example 4.2* —

Figure 4.6a shows the phase calculated within only two iterations. An optical element with such a phase has the Bessel spectrum shown in Fig. 4.6b. It is seen from Fig. 4.6b that the sum in Eq. (4.26) is effectively contributed by only four terms with the roots γ_{15} to γ_{18}. The light beam generated by such an optical element will propagate almost without diffraction. Figure 4.7 depicts the radial distribution of the intensity of this beam at different distances: $z = 100$ mm, 150 mm, and 170 mm. Parameters of the calculation are $\beta = 50\,\mathrm{mm}^{-1}$, $R = 1$ mm, $k = 10^4\,\mathrm{mm}^{-1}$, the number of pixels is 512, and the discreteness along the variable r is 0.02 mm. With these parameters, the mode effective radius, i.e. the distance from the axis to the first zero, is approximately 48 μm (see Fig. 4.7). Curve 4 (Fig. 4.7) shows the radial intensity of the Gaussian beam for $z = 150$ mm, whose radius was equal to 48 μm for $z = 0$.

It is noteworthy that the conic axicon [87] also represents a phase element forming a zero-order quasimode. This can be deduced from the expansion

$$\mathrm{e}^{-i\beta r} = \sum_{n=-\infty}^{\infty} i^n J_n(\beta r) \tag{4.32}$$

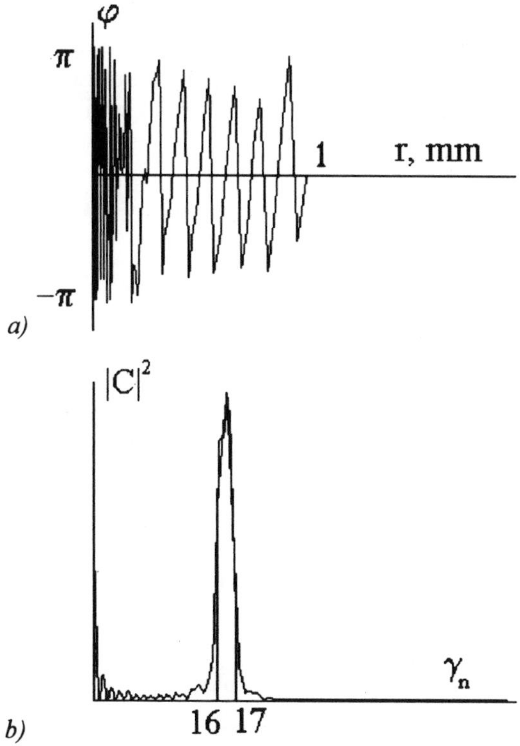

Figure 4.6 (a) radial DOE phase; (b) intensity in the Bessel spectrum

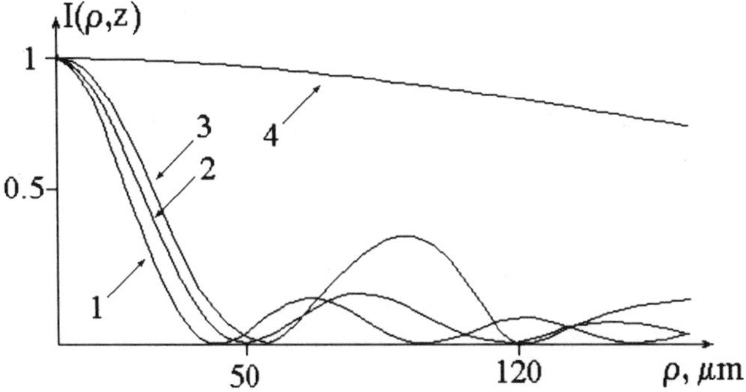

Figure 4.7 Radial intensity distribution generated by the DOE (Fig. 4.6a) at different distances: $z = 100$ mm (1), 150 mm (2), 170 mm (3); curve 4 shows the radial intensity of the Gaussian beam for $z = 150$ mm

where only the term $J_0(\beta r)$ will effectively contribute to the sum, whereas the contribution of the other terms will be small because of the diffraction (divergence) of the beams characterized by the amplitude $J_n(\beta r)$, at $|n|>0$. Note also that the phase illustrated in Fig. 4.6a is similar to the linear phase of the axicon of Eq. (4.32).

4.3 DOEs for the Formation of Gaussian Modes

References [56,104–106] report studies of phase optical elements (modans) which are matched to the Gaussian modes of laser radiation and generate the light beams as a superposition of Gauss–Hermite modes or Gauss–Laguerre modes. In the subsequent text we deal with iterative algorithms for calculating such optical elements.

4.3.1 *Algorithms for Calculating the Formers of Gauss–Hermite Modes*

It is easy to check that there exists a partial solution for the paraxial equation in a 1D case

$$2ik\frac{\partial E(x,z)}{\partial z}+\frac{\partial^2 E(x,z)}{\partial x^2}=0 \tag{4.33}$$

where z is an axis along the beam propagation, which can be represented by

$$E_n(x,z)=a^{-n-1}\mathrm{e}^{-(x/a)^2}H_n(x/a) \tag{4.34}$$

$a^2(z)=a_0^2+2iz/k$, a_0^2 is the beam radius at $z=0$, and $H_n(x)$ is a Hermite polynomial.

Therefore, the general solution to Eq. (4.33) can be written as a linear combination of its partial solutions given by Eq. (4.34)

$$E(x,z)=\mathrm{e}^{-(x/a)^2}\sum_{n=0}^{\infty}C_n a^{-n-1}H_n(x/a) \tag{4.35}$$

In order for the light field with the amplitude in the form of Eq. (4.35) to propagate along the z-axis, one should form in plane $z=0$ the following complex amplitude

$$E_0(x)=\mathrm{e}^{-(x/a_0)^2}\sum_{n=0}^{\infty}C_n a_0^{-n-1}H_n(x/a_0) \tag{4.36}$$

Thus, the problem of calculating phase optical elements producing Hermite beams with an arbitrary modal composition can be stated as follows [103]. It is necessary to find a phase $\varphi(x)$ satisfying the set of $N+1$ algebraic equations

$$|C_n| = \left| \int_{-\infty}^{\infty} A_0(x)\,\mathrm{e}^{i\varphi(x)} P_n(x)\,\mathrm{d}x \right|, \quad n = \overline{0, N} \tag{4.37}$$

where

$$A_0(x)\,\mathrm{e}^{i\varphi(x)} = \sum_{n=0}^{N} C_n P_n(x) \tag{4.38}$$

$$P_n(x) = (2^n n! \sqrt{n})^{-1/2} \mathrm{e}^{-(x^2/2)} H_n(x) \tag{4.39}$$

$P_n(x)$ is the Hermite function and A_0 is the illuminating beam amplitude. The modules of the C_n coefficients are pregiven and specify the contribution of each Hermite mode to the total beam.

The Hermite function in Eq. (4.39) obeys the orthogonality relationship

$$\int_{-\infty}^{\infty} P_n(x) P_m(x)\,\mathrm{d}x = \delta_{mn} \tag{4.40}$$

The algorithm for searching for the solution to the set of Eqs (4.37) is analogous to the iterative Gerchberg–Saxton algorithm [14]. The initial estimate $\varphi_0(x)$ of the phase is chosen to be stochastic. $N+1$ coefficients C_n are then calculated using Eqs (4.37). The $C_n^{(k)}$ coefficients derived in the kth iteration step are subject to the substitution

$$\bar{C}_n^{(k)} = B_n \frac{C_n^{(k)}}{|C_n^{(k)}|} \tag{4.41}$$

where $B_n \geq 0$ are the pregiven numbers characterizing the distribution of light energy between the modes. Next, the sum of Eq. (4.38) is computed, whose argument turns out to be a new estimate $\varphi_k(x)$ of the desired phase, and so forth.

The proof of the convergence of the present algorithm is analogous to that given in Appendix E.

One should note that all the above reasoning can be completely extended to a 2D case if one needs to obtain the phase $\varphi(x, y)$ of an optical element forming the light beam consisting of a pregiven set of Hermite beams with desired energy weights

$$A_0(x, y)\,\mathrm{e}^{i\varphi(x,y)} = \sum_{n=0}^{N} \sum_{m=0}^{M} C_{mn} P_n(x) P_m(y) \tag{4.42}$$

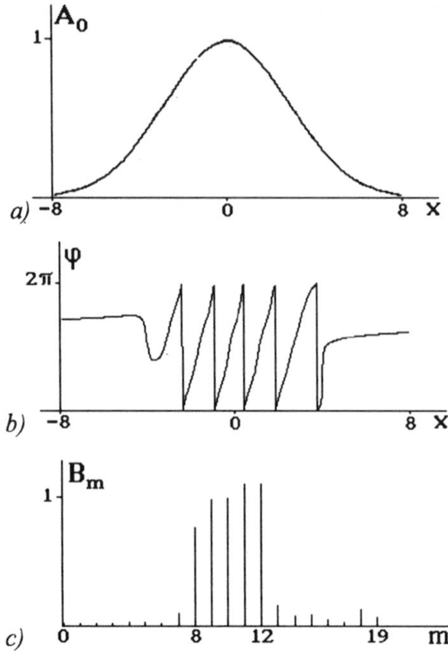

Figure 4.8 Calculation of the Hermite beam: (a) the illuminating beam amplitude; (b) the DOE phase; (c) modules of the coefficients C_k

— *Example 4.3*

Upon numerical simulation, pixels along the x-variable were taken in 0.02 mm steps. An integration of Eq. (4.37) in infinite terms was replaced by the integration over an interval $[-a, a]$, $a = 6$ mm. With increasing a the C_n coefficients varied by 0.1%. The amplitude $A_0(x)$ of the illuminating beam was chosen to be the Gaussian one (see Fig. 4.8a). The number of terms N in the sum in Eq. (4.38) was equal to 20.

Figure 4.8b depicts the phase of an optical element which is illuminated by the plane beam with a Gaussian intensity distribution (see Fig. 4.8a) and generates the Hermite beam that effectively represents a superposition of five Gauss–Hermite modes with constant modules of the coefficients C_k, $k = 8, 12$. The modules of the coefficients are shown in Fig. 4.8c. Note that the overall inclination of the optical element phase (see Fig. 4.8b) resulting in the axial displacement of the Hermite beam upon its propagation arises because there exist modes with odd numbers in the sum of Eq. (4.38).

Figure 4.9 shows the r.m.s. deviations for the amplitude in the optical element plane (curve 1)

$$\delta_A = \left[\int_{-a}^{a} |A_{k+1} - A_0|^2 \mathrm{d}x \right]^{1/2} \left[\int_{-a}^{a} A_0^2 \mathrm{d}x \right]^{-1/2} \tag{4.43}$$

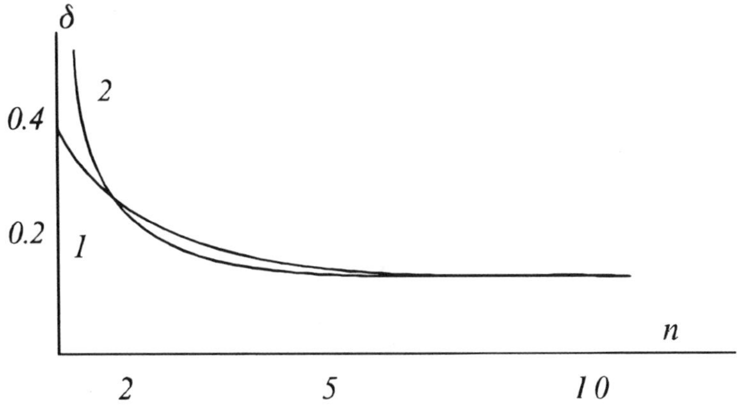

Figure 4.9 Plots of deviations (see text)

and for the coefficients of the series in Eq. (4.38) (curve 2)

$$\delta_C = \left[\sum_{n=0}^{N} (|C_n^{(k)}| - B_n)^2 \right]^{1/2} \left[\sum_{n=0}^{N} B_n^2 \right]^{-1/2} \tag{4.44}$$

versus the number of iterations.

From Fig. 4.9, the deviations are seen to decrease monotonically with increasing number of iterations. Note that the efficiency of representation of the Hermite beam via the five modes amounts in this case to 90% (see Fig. 4.8c). The efficiency can be found from the relation

$$E = \left[\sum_{\Omega} |C_n^{(k)}|^2 \right] \left[\sum_{n=0}^{N} |C_n^{(k)}|^2 \right]^{-1} \tag{4.45}$$

where k is the number of the iteration and Ω is the set of the numbers that produce non-zero coefficients.

Reference [106] deals with optical elements that have been given the name 'multichannel modans'. They consist of a set of Gaussian modes, each of which has its own carrier frequency, and are used for branching the illuminating beam onto several beams (channels) capable of exciting or analyzing several modes in parallel. These elements are well suited for the parallel entering of laser light into a set of light fibres.

In what follows, we deal with the problem of calculating the phase of the above elements [31,107]. This problem is somewhat different from the preceding one [103] because the present case involves the calculation of a phase optical element forming Hermite beams in desired diffraction orders and with a desired distribution of the light energy between these orders, i.e. in this case each light mode propagates at its own angle to the optical axis.

Therefore, to find the DOE phase $\varphi(x)$, one should use instead of Eq. (4.38) the following relation

$$A_0(x)\mathrm{e}^{i\varphi(x)} = \sum_{n=0}^{N} C_n P_n(x)\mathrm{e}^{-i\alpha_n x} \tag{4.46}$$

where $P_n(x)$ is the Hermite function specified in Eq. (4.39) and α_n is the carrier spatial frequency: $\alpha_1 < \alpha_2 < \ldots < \alpha_N$. The modules of the coefficients C_n are, as before, fixed values, with their arguments being free parameters of the task. The Hermite functions are eigenfunctions for the Fourier transform (see [108], p. 180)

$$\int_{-\infty}^{\infty} P_n(x)\mathrm{e}^{-i\alpha x}\mathrm{d}x = (-i)^n P_n(\alpha) \tag{4.47}$$

Making use of Eq. (4.47), we obtain the expression

$$\int_{-\infty}^{\infty} P_n(x)P_m(x)\exp[-i(\alpha_n - \alpha_m)x]\,\mathrm{d}x = i^{n+m} \times \int_{-\infty}^{\infty} P_n(\beta)P_m(\beta + \alpha_n - \alpha_m)\,\mathrm{d}\beta \tag{4.48}$$

The function on the right-hand side of Eq. (4.48) is a correlation of two Hermite functions and has the effective domain of argument beyond which we can neglect the function values twice as great as the effective domain of argument of the Hermite function of higher order. Therefore, by choosing the difference between two successive spatial carrier frequencies $|\alpha_{n+1} - \alpha_n|$ to be large enough (in practice, this implies that the generated Hermite beams should be separated with a margin), we can consider the functions

$$P_n(x)\exp[-i\alpha_n x]$$

as being approximately orthogonal and deduce the coefficients of the sum in Eq. (4.46) from the relationship

$$C_n = \int_{-\infty}^{\infty} A_0(x)\mathrm{e}^{i\varphi(x)+i\alpha_n x} P_n(x)\,\mathrm{d}x, \quad n = \overline{0, N} \tag{4.49}$$

Using Eqs (4.46) and (4.49), we can organize an iterative procedure of searching for the phase $\varphi(x)$ of the optical element, which would be analogous to the algorithm described by Eqs (4.37)–(4.39).

The C_n coefficients derived from Eq. (4.49) in the kth iteration step are subject to the following replacements: their modules are replaced by pregiven numbers B_n, with the phases (arguments) remaining unchanged. Next, the sum in Eq. (4.46) is found and the resulting function undergoes a similar replacement: its module is replaced by the amplitude $A_0(x)$ of the illuminating beam, whereas the phase remains unchanged and represents a new estimate of the desired phase.

It is apparent that the foregoing reasoning can be applied to a 2D situation if, instead of Eq. (4.46), we employ the relation

$$A_0(x,y)\,\mathrm{e}^{i\varphi(x,y)} = \sum_{n=0}^{N}\sum_{m=0}^{M} C_{nm} P_n(x) P_m(y)\,\mathrm{e}^{-i(\alpha_n x+\beta_m y)} \tag{4.50}$$

— *Example 4.4* —

For a particular calculation we have chosen the following parameters. The variables x and y ranged over the interval $[-10, 10]$ with a step of 0.02. The total number of pixels was 128×128. The number of terms in the sum in Eq. (4.50) was equal to $M = N = 9$. The amplitude $A_0(x, y)$ of the illuminating beam was chosen to be constant.

Figure 4.10 shows the phase of a four-channel optical element that has been calculated during eight iterations and is able to form in the lens focal plane the intensity distribution, and the phase distribution. Figure 4.10b shows: mode $(0, 1)$ at the upper left corner, mode $(1, 1)$ at the upper right corner, mode $(1, 2)$ at the bottom left, and mode $(2, 0)$ at the bottom right. About 95% of the energy of the illuminating beam goes to form these four Gaussian modes, with the energy to be distributed in equal proportions between the modes with an average error of 2%.

The mean deviation of the intensity of the produced modes from the ideal ones is 21%. Figure 4.11 shows how the error varies with iterations. The error is seen to decrease monotonically. An increase in the convergence rate after the sixth iteration (see Fig. 4.11) is due to the employment of an adaptive–iterative algorithm [26]. Figure 4.10c depicts the phase in the focal plane. It is seen that in the areas of localization of the four modes the phase has characteristic jumps of π in passing over the zero lines of intensity.

An optical element with the phase shown in Fig. 4.10a can be utilized not only for the formation of Gauss–Hermite modes propagating in space in various directions, but also as a spatial filter matched to the foregoing four modes.

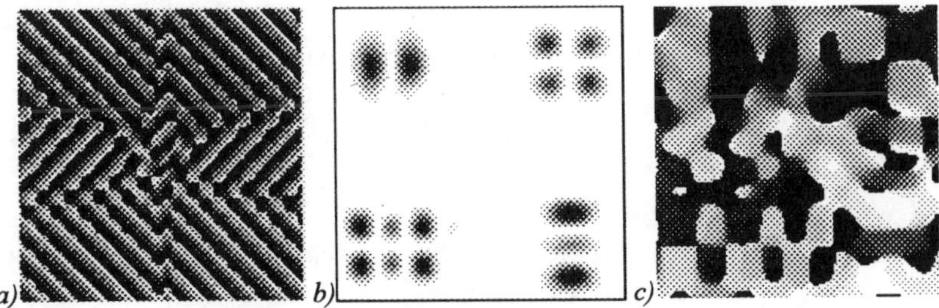

Figure 4.10 Calculation of a four-channel former of Hermite modes: (a) the DOE phase; (b) the intensity; (c) the phase distributions in the focal plane

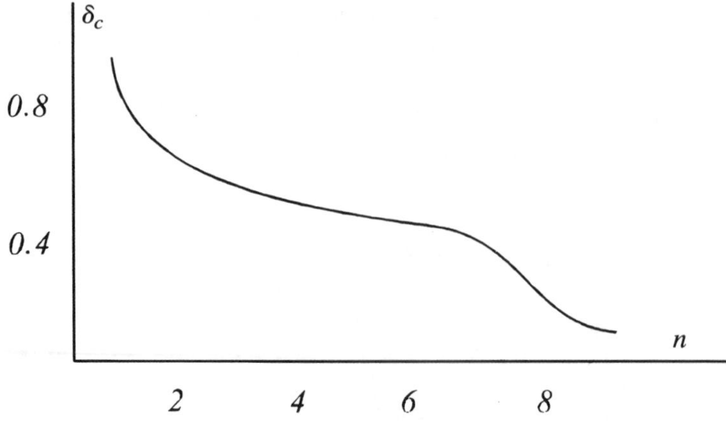

Figure 4.11 Error (Eq. (4.44)) vs number of iterations

Actually, if such an optical element is illuminated by the light beam with an amplitude proportional to the amplitude of a certain mode $T_{mn} = P_n(x)P_m(y)$, the correlation intensity peak will appear in the frequency plane at point (α_n, β_m). Figure 4.12 illustrates intensity distributions in the frequency plane if the optical element (Fig. 4.10a) is illuminated by mode T_{01}, mode T_{20}, mode $T_{11} + T_{12} + T_{20}$ or mode T_{02}. Note that the last mode was absent when the DOE was calculated.

4.3.2 Algorithm for Calculating Formers of Gauss–Laguerre Modes

In this section we discuss the problem of calculating phase optical elements that can shape axial light beams as a result of them being a set of Gauss–Laguerre modes with a preset energy distribution between the modes [31]. In addition, we treat the problem of calculating multichannel optical elements serving to form Laguerre modes in desired diffraction orders. The line of reasoning will, in many respects, follow that in the previous section.

Paraxial equation (4.33) in cylindrial coordinates will read [109]

$$2ik\frac{\partial E(r,\theta,z)}{\partial z} + \frac{1}{r}\frac{\partial}{\partial r}\left[r\frac{\partial E(r,\theta,z)}{\partial r}\right] + \frac{1}{r^2}\frac{\partial^2 E(r,\theta,z)}{\partial \theta^2} = 0 \qquad (4.51)$$

where $x = r\cos\varphi$, $y = r\sin\varphi$, and $z = z$.

For axial beams $E(x,y,z) = E(r,z)$, thus instead of Eq. (4.51) we obtain the relation

$$2ik\frac{\partial E(r,z)}{\partial z} + \frac{\partial^2 E(r,z)}{\partial r^2} + \frac{1}{r}\frac{\partial E(r,z)}{\partial r} = 0 \qquad (4.52)$$

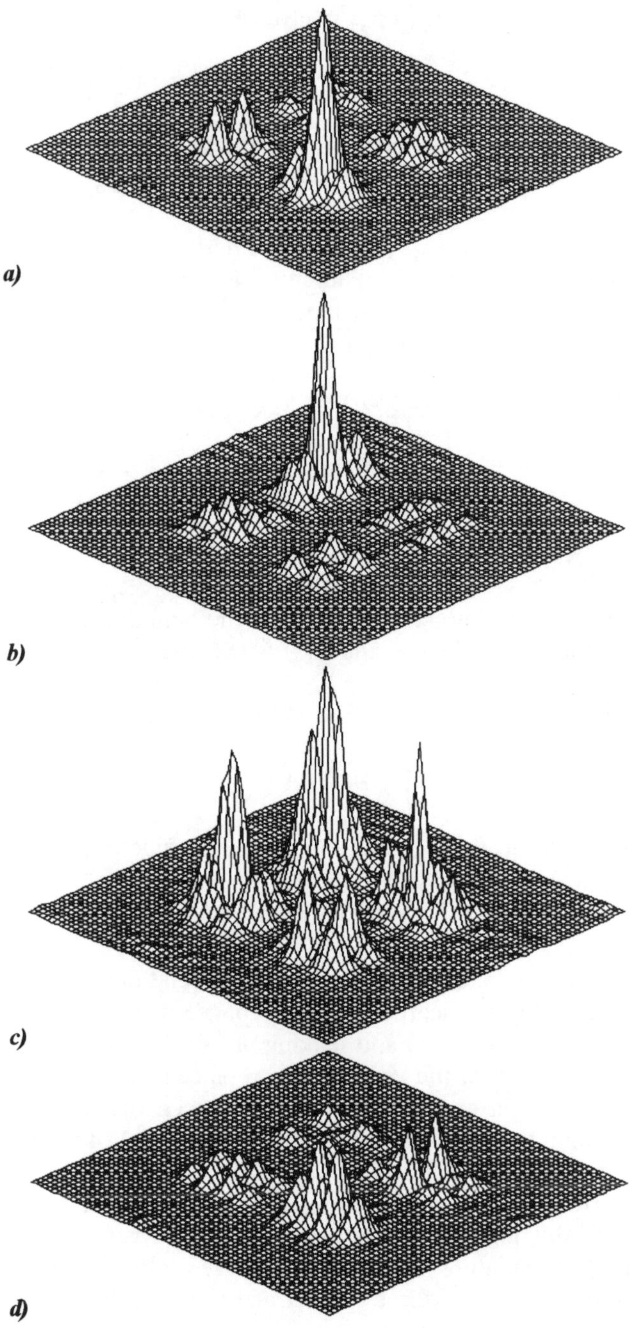

Figure 4.12 The intensity distributions in the frequency plane if the DOE (Fig. 4.10a) is illuminated by: (a) mode T_{01}; (b) mode T_{20}; (c) mode $T_{11}+T_{12}+T_{20}$; (d) mode T_{02}

A partial solution to Eq. (4.52) is given by

$$E_m(r, z) = (1 + iD)^{-m-1} e^{-(r/a)^2} L_m\left(\frac{r^2}{a^2}\right) \tag{4.53}$$

where $a^2(z) = a_0^2(1 + iD)$, $D = 2z/ka_0^2$ is the Fresnel parameter, a_0 is the beam effective radius at z equal to zero, and $L_m(x)$ is the Laguerre polynomial [110].

The general solution to Eq. (4.52) can be derived as a linear combination of partial solutions of Eq. (4.53). Therefore, to shape a light beam consisting of a pregiven set of Gauss–Laguerre modes, one should find the light complex amplitude at $z = 0$ in the form

$$A_0(r) e^{i\varphi(r)} = \sum_{n=0}^{N} C_n S_n(r) \tag{4.54}$$

where

$$S_n(r) = \frac{\sqrt{(2)} e^{-r^2/2}}{n!} L_n(r^2) \tag{4.55}$$

$S_n(r)$ is the Laguerre function, $A_0(r)$ is the amplitude of the radially symmetrical illuminating beam, and $\varphi(r)$ is the sought radially symmetrical DOE phase.

The coefficients of the sum in Eq. (4.54) are found from

$$C_n = \int_0^\infty A_0(r) e^{i\varphi(r)} S_n(r) r \,\mathrm{d}r \tag{4.56}$$

which is based on the orthogonality of Laguerre functions

$$\int_0^\infty S_n(r) S_m(r) r \,\mathrm{d}r = \delta_{nm} \tag{4.57}$$

Since relations (4.54) and (4.56) are similar to relations (4.38) and (4.37), an iterative algorithm for searching for the phase will be analogous. Given a guess of desired phase $\varphi_k(r)$ and making use of Eq. (4.56) we deduce the coefficients $C_n^{(k)}$ in which the amplitude is replaced by the preset values B_n characterizing the light energy distribution between the modes, with arguments remaining unchanged. The next estimate $\varphi_{k+1}(r)$ of the phase will be equal to the argument of a function that will be a result of summation using Eq. (4.54) with altered coefficients, and so forth. The proof of the convergence of this algorithm is analogous to that given in Appendix E.

The task of calculating multichannel phase optical elements forming Gauss–Laguerre modes in desired diffraction orders can be posed in a similar way, based on the expression

$$A_0(x, y) e^{i\varphi(x,y)} = \sum_{n=0}^{N} C_n S_n[\sqrt{(x^2 + y^2)}] \exp[-i(\alpha_n x + \beta_n y)] \tag{4.58}$$

where (α_n, β_n) is the spatial carrier frequency vector specifying the nth diffraction order.

In the subsequent discussion it is convenient to express Eq. (4.58) in polar coordinates

$$A_0(r, \theta)\,\mathrm{e}^{i\varphi(r,\theta)} = \sum_{n=0}^{N} C_n S_n(r) \exp[-ir\rho_n \cos(\theta - \theta_n)] \tag{4.59}$$

where (ρ_n, θ_n) is the spatial carrier frequency vector in polar coordinates.

The Laguerre function in Eq. (4.55) is an eigenfunction of the Hankel transform of zero order

$$\int_0^\infty S_n(r) J_0(r\varphi)\, r\,\mathrm{d}r = (-1)^n S_n(\rho) \tag{4.60}$$

This can be easily shown with the help of a tabulated integral ([102], p. 475)

$$\int_0^\infty x^{\lambda/2} \mathrm{e}^{-cx/2} J_\lambda(b\sqrt{x})\, L_n^\lambda(cx)\,\mathrm{d}x = \frac{(-1)^n 2b^\lambda}{c^{\lambda+1}} \mathrm{e}^{-(b^2/2c)} L_n^\lambda\left(\frac{b^2}{c}\right) \tag{4.61}$$

where $L_n^\lambda(x)$ is the adjoin (or generalized) Laguerre polynomial and $J_\lambda(x)$ is the λth-order Bessel function.

To be able to derive the coefficients of the sum in Eq. (4.59) from the relation

$$C_n = \int_0^\infty \int_0^{2\pi} \underset{\theta}{A_0}(r, \theta)\,\mathrm{e}^{i\varphi(r,\theta)} S_n(r)\,\mathrm{e}^{ir\rho_n \cos(\theta-\theta_n)}\, r\,\mathrm{d}r\,\mathrm{d}\gtrsim \tag{4.62}$$

we show that for a sufficiently large spacing between the neighbouring vectors of spatial frequencies

$$w_{n,n+1} = \sqrt{[\rho_n^2 + \rho_{n+1}^2 - 2\rho_n\rho_{n+1}\cos(\theta_n - \theta_{n+1})]} \tag{4.63}$$

the function $S_n(r)\exp[ir\rho_n\cos(\theta - \theta_n)]$ may be considered as being nearly orthogonal. This will be the case if the following integral is small

$$I_{mn} = \int_0^\infty \int_0^{2\pi} S_m(r) S_n(r) \exp[ir\rho_m \cos(\theta - \theta_m) - ir\rho_n \cos(\theta - \theta_n)]\, r\,\mathrm{d}r\,\mathrm{d}\theta \tag{4.64}$$

On integration with respect to angle θ, Eq. (4.64) can be represented as

$$I_{mn} = 2\pi \int_0^\infty S_m(r) S_n(r) J_0(w_{mn} r)\, r\,\mathrm{d}r \tag{4.65}$$

where w_{mn} is the distance between the two spatial frequency vectors specified in Eq. (4.63).

Using a tabulated integral ([102], p. 475)

$$\int_0^\infty x^{(\gamma+\lambda)/2} e^{-cx/2} J_{\lambda+\gamma}(b\sqrt{x}) L_m^\gamma(cx) L_n^\lambda(cx)\,dx$$
$$= \frac{(-1)^{m+n}}{c^{\gamma+\lambda+1}} \left[\frac{b}{2}\right]^{\gamma+\lambda} \exp\left[-\frac{b^2}{4c}\right] L_m^{\gamma+m-n}\left(\frac{b^2}{4c}\right) L_n^{\lambda-m+n}\left(\frac{b^2}{4c}\right) \tag{4.66}$$

where

$$L_n^\lambda(x) = \frac{x^{-\lambda} e^x}{n!} \frac{d^n}{dx^n}[x^{n+\lambda} e^{-x}] \tag{4.67}$$

is the adjoin Laguerre polynomial, instead of Eq. (4.65) we obtain

$$I_{mn} = \pi(-1)^{m+n} \exp\left[\frac{-w_{mn}^2}{4}\right] L_m^{m-n}\left(\frac{w_{mn}^2}{4}\right) L_n^{n-m}\left(\frac{w_{mn}^2}{4}\right) \tag{4.68}$$

It is seen from Eq. (4.68) that with w_{mn} tending to infinity, I_{mn} drops exponentially, which means that at a sufficiently large distance between two successive vectors of spatial frequency the values of Eq. (4.65) will be close to zero and we can employ Eq. (4.62).

The iterative procedure of calculating the phase optical elements forming Laguerre modes in desired diffraction orders is much the same as that in the previous section. If there exists an estimate $\varphi_k(x, y)$ of the desired phase at the kth iteration step, then the $C_n^{(k)}$ coefficients are derived from Eq. (4.62) and their module is replaced by the preset number B_n, whereas the argument remains unchanged.

These coefficients are utilized in Eq. (4.59) to deduce the sum whose argument will be a new estimate $\varphi_{k+1}(x, y)$ of the desired DOE phase.

— *Example 4.5*

Figure 4.13a depicts the phase of an optical element calculated within ten iterations using Eqs (4.59) and (4.62). An optical element with such a phase

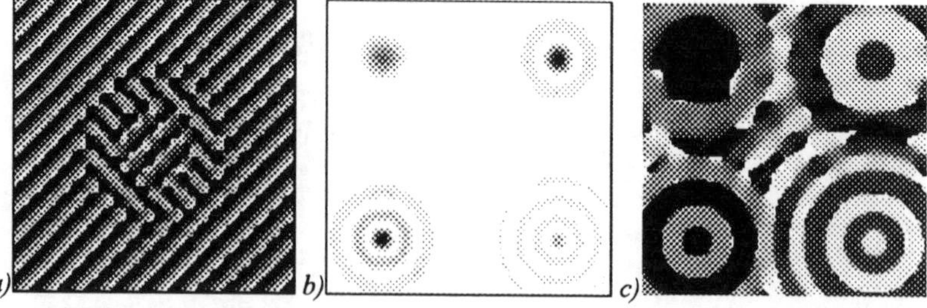

Figure 4.13 Calculation of a four-channel former of Gauss–Laguerre modes: (a) the DOE phase; (b) the intensity; (c) phase distributions in the focal plane

transforms the illuminating Gaussian beam into the first four Gauss–Laguerre modes with a 90% efficiency in the lens focal plane. The modes' intensities that have a separating distance and equal energies are shown in Fig. 4.13b.

5

Design of Multiorder Diffraction Gratings with a Pregiven Intensity of Diffraction Orders

5.1 Gradient Method with Analytical Initial Approximation for Multiorder Binary Gratings Design

Multiorder gratings or fan-out gratings, as they are usually called, are thin periodic phase relief structures which can generate a 1D or 2D array of light beams. Their use has been reported in many applications such as multiple imaging arrays, coherent addition of laser beams, fibre optic star couplers, free-space optical interconnects, optical processing, optical computing and optical communication [111–116].

Many approaches have been used to design multiorder binary as well as multilevel and continuous-relief diffraction gratings. Among these are stochastic techniques (direct binary search [59,117], iterative discrete on-axis algorithms [45], simulated annealing algorithms [118–121]), Gerchberg–Saxton phase retrieval algorithms [14,19,42,122] and gradient methods [20,40,123].

Binary gratings are of particular interest because of the ease of their manufacture. Solving a system of N nonlinear equations is required in order to design a binary grating generating $2N+1$ orders [41,124]. For asymmetrical grating profiles, finding a solution requires more effort than in the symmetric case; on the other hand, solutions with larger minimum feature size and higher efficiency can be found [41,124].

For applications such as laser beam homogenizers and shapers as well as fibre optic networks it is of real value to extend the binary grating design to a large number of orders; when the number of orders increases, the random search for the binary grating profile is time-consuming and leads to nonstable convergence, the solution of the nonlinear equations system is numerically prohibitive and gradient procedures with a random starting-point

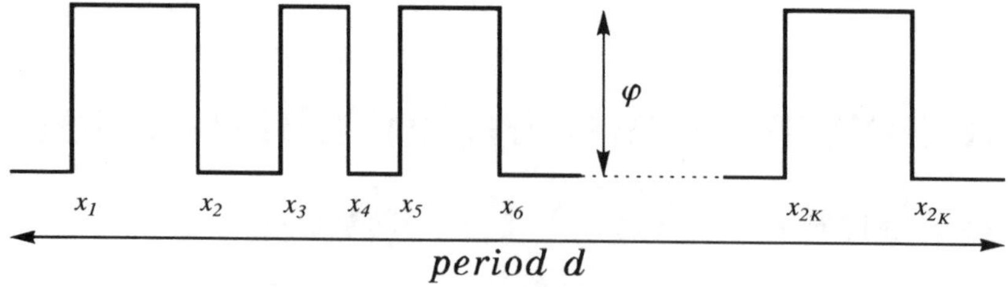

Figure 5.1 Phase profile of a binary grating period

lead to the local minima of merit function, which do not represent the solution of the problem. In this section, to calculate multiorder asymmetrical binary gratings we consider an initial approximation of the grating profile based on the properties of a 1D element focusing into a line. The kind of initial approximation chosen allows a quick convergence of the iteration procedure. When the method is applied to the calculation of gratings of any order up to 281, it gives energy efficiencies of 78–84% and root mean square (r.m.s.) errors of 1–5% within a few minutes on a 386 computer.

5.1.1 The Initial Approximation of the Binary Grating Profile

It is known that the phase profile of a binary grating period consists of K grooves of equal depth but different width (Fig. 5.1). The coordinates $x_1, \ldots, x_{2K}$ of the grooves represent phase transitions with amplitude φ. The position of the coordinates and the depth of the grooves are the parameters that influence the intensity of diffraction orders.

We define the minimum feature size as the minimum distance between adjacent coordinates, and the fill factor of the grating as the ratio between the part of the profile with phase φ and the overall period. It has been demonstrated that, in general, the error of the scalar theory is significant (>5%) when the minimum feature size is less than 14λ [125]; however, at the optimal depth, even for feature sizes of $\sim 2\lambda$ the error is minimized when the fill factor approaches 50% [125]. As is shown in [125], for binary structures the scalar accuracy is not very sensitive to variations of the incidence angle: without any restriction we assume unit period and normal incidence in the calculation. Appendices F and G deal with the algorithms for synthesis of DOEs with regard to the electromagnetic radiation polarization.

With the purpose of determining a suitable approximation to the binary profile, let us introduce a 1D phase DOE focusing into a line with intensity distribution $I(\xi)$ along the line. Let d_1 be the dimension of a focal line located at a distance f (see Fig. 5.2). The focal line can be considered as a series of adjacent diffraction spots of size $\Delta = \lambda f/d$, where λ is the wavelength and

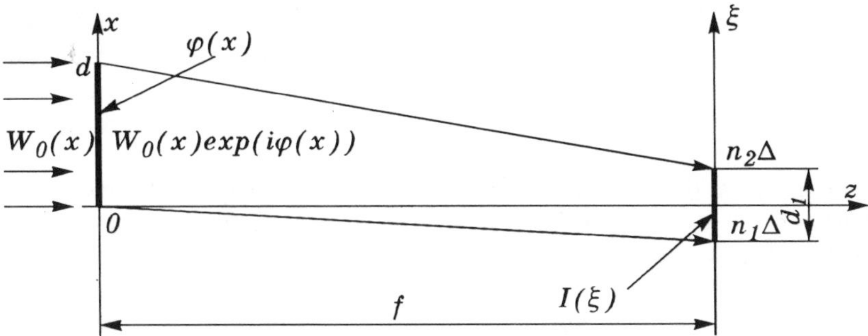

Figure 5.2 Geometry of focusing into a line

d is the dimension of the 1D DOE. Let $I(\xi)$ be the intensity along the line defined in the interval $(n_1\Delta, n_2\Delta)$; we have $d_1 = (n_2 - n_1)\Delta$. For the convergent cylindrical incident beam with the complex amplitude

$$W_0(x) = \exp\left(-\frac{ikx^2}{2f}\right), \quad k = \frac{2\pi}{\lambda}$$

the phase function $\varphi(x)(0 \le x \le d)$ of the 1D DOE computed with a ray-tracing approach is given by the equation [37,126]

$$\varphi(x) = \frac{2\pi}{d}(n_2 - n_1)\int_0^x \kappa(\alpha)\,\mathrm{d}\alpha + \frac{2\pi}{d}n_1 x \tag{5.1}$$

where $\kappa(\alpha)$ is the solution of the differential equation

$$\frac{\mathrm{d}\kappa(\alpha)}{\mathrm{d}\alpha} = \left[I\left(\frac{\kappa(\alpha) - n_1\Delta}{(n_2 - n_1)\Delta}\right)\right]^{-1} \tag{5.2}$$

with boundary conditions $\kappa(0) = 0$, and $\kappa(1) = 1$. Following the terminology used in [37,126], we call the DOE of Eqs (5.1) and (5.2) the focusator.

The function $\varphi(x)$ represents the phase function of a nonbinary diffraction grating originating $(n_2 - n_1 + 1)$ diffraction orders. From the Fresnel–Kirchhoff integral for $\xi = \xi_j = j\Delta$, the intensities in the focal plane of the focusator

$$I(\xi_j) = \frac{1}{\lambda f}\left|\int_0^d \exp\left(i\varphi(x) - i\frac{2\pi j}{d}x\right)\mathrm{d}x\right|^2 \tag{5.3}$$

are proportional to the squared modules of the Fourier coefficients of the function $\exp(i\varphi(x))$. We can then consider $\varphi(x)$ as the phase function of a nonbinary diffraction grating with a period of d and order intensities given by

$$I_j = \begin{cases} I(j\Delta), & n_1 \le j \le n_2 \\ 0, & \text{else} \end{cases} \tag{5.4}$$

To calculate the binary grating generating $2N+1$ orders in the interval $[-N\Delta, +N\Delta]$, we suggest defining a binary phase function of the form [127,128]

$$\varphi_b(x) = \Phi[\overline{\varphi}(x)] \tag{5.5}$$

where $\overline{\varphi}(x)$ is the nonbinary phase function of Eq. (5.1) modulo 2π generating $(N+1)$ orders in the interval $[0, N\Delta]$, and $\Phi[\eta]$ is the phase function of a two-order binary grating given by

$$\Phi[\eta] = \begin{cases} 0, & \eta \in [0, \pi] \\ \pi, & \eta \in [\pi, 2\pi] \end{cases} \tag{5.6}$$

and describing a point-to-point nonlinear transformation of the phase function $\overline{\varphi}(x)$. To understand the operation of the binary grating $\varphi_b(x)$ in the interval $[0, 2\pi]$ we expand the function $\exp(i\Phi[\eta])$ into a Fourier series

$$\exp(i\Phi[\eta]) = \sum_{n=-\infty}^{\infty} c_n \exp(in\eta) \tag{5.7}$$

where

$$c_n = \begin{cases} \dfrac{1-(-1)^n}{i\pi n}, & n \neq 0, n = \pm 1, \pm 2, \ldots \\ 0, & n = 0 \end{cases} \tag{5.8}$$

are the Fourier coefficients of the function $\exp(i\Phi[\eta])$. Substituting $\eta = \overline{\varphi}(x)$ into Eq. (5.7), the complex transmission function of binary grating $\varphi_b(x)$ with respect to 2π phase periodicity can be written in the form

$$\exp(i\Phi[\overline{\varphi}(x)]) = \sum_{n=-\infty}^{\infty} c_n \exp(in\varphi(x)) \tag{5.9}$$

which can be read as a superposition of beams with the phase functions

$$\varphi_n(x) = n\varphi(x) \tag{5.10}$$

Since $|c_1|^2 = |c_{-1}|^2 = 0.405$, we have 81% of the illuminating beam power in the -1 and $+1$ diffraction orders. The images generated in these two main orders correspond to the phase functions $-\varphi(x)$ and $+\varphi(x)$. These are, respectively, the phase functions of two nonbinary gratings generating the orders in the intervals $[-N\Delta, 0]$ and $[0, N\Delta]$, whose summation gives the desired number of orders in the interval $[-N\Delta, N\Delta]$. Following the previous considerations (see Eqs (5.1)–(5.4)), the binary structure of Eq. (5.5) can be considered as an approximated solution of the binary grating profile. The function $\varphi(x)$ was calculated by the ray-tracing method; in any case, we only considered the two main diffraction orders without regard for the interference between them. This is why – even if the efficiency associated with the approximated solution is close to 70–80% – the r.m.s. error remains in the range of 35–40%; we then use this approximated solution only as the starting-point of the optimization process.

According to Eqs (5.1)–(5.6), the coordinates of the initial profile of a grating with equal intensity in $2N+1$ diffraction orders are defined by the relation

$$x_i = \sqrt{\left(\frac{i}{N}\right)}, \quad i = \overline{1, N} \tag{5.11}$$

Similarly, for a grating with symmetrical linear distribution of the intensity in diffraction orders

$$I_i = I_0 + c|i|, \quad i = \overline{-N, N}$$

the coordinates of the initial grating profile are defined by the relation

$$\frac{2N}{(b-1)(b^2-1)} \times \{-3x_i(b^2-1) - 2 + 2[1 + (b^2-1)x_i]^{3/2}\} = i, \quad i = 1, 2, \ldots \tag{5.12}$$

where $b = 1 + cN/I_0$.

5.1.2 Optimization Procedure

With the purpose of optimizing the grating structure, we define a merit function that represents the difference between the desired intensities $\hat{I}_j$ in orders and the calculated ones I_j in the form

$$\varepsilon(\mathbf{x}) = \sum_{j=-N}^{N} (I_j(\mathbf{x}) - \hat{I}_j)^2 \tag{5.13}$$

where the vector $\mathbf{x} = (x_1, \ldots, x_{2K})$ represents the coordinates of the grooves. The assumption of unitary period allows the representation of the squared module of Fourier coefficients in the following form [45,124]

$$\begin{cases} I_j(\mathbf{x}) = \sin^2(\varphi/2)\,\dfrac{(C_j^2 + S_j^2)}{(\pi j)^2}, & j \neq 0, j = \pm 1, \pm 2, \ldots \\ I_0(\mathbf{x}) = 1 - 4Q(1-Q)\sin^2(\varphi/2) \end{cases} \tag{5.14}$$

where

$$C_j = \sum_{i=1}^{2K} (-1)^i \cos(2\pi j x_i)$$

$$S_j = \sum_{i=1}^{2K} (-1)^i \sin(2\pi j x_i) \tag{5.15}$$

$$Q = \sum_{i=1}^{2K} (-1)^i x_i$$

Since the squared modules of Fourier coefficients are proportional to the intensity in the diffraction orders, we replace the squared modules with the intensities for convenience. Analysis of Eqs (5.14) and (5.15) shows that $I_j = I_{-j}$. Due to that symmetry, the calculation of a $2N+1$ binary grating with an asymmetrical profile requires a minimum of $N+1$ free parameters; accordingly, the number of grooves needed must be $K \geq N/2$ [45,124].

The calculation method consists of a gradient minimization of $\varepsilon(\mathbf{x})$ searching for the best coordinates of the grooves; the gradient search is then the iterative correction of the grooves' coordinates. In the general case, the main problem associated with gradient procedures is the choice of an initial profile to obtain a stable convergence. As a rule, with a random generation of the initial profile, the algorithm stagnates with the r.m.s. deviation of the final order intensities from the designed ones in the range of 70–80%. With the initial analytical approximation we propose, we managed to obtain quick and stable convergence with high efficiency and low r.m.s. error using different gradient procedures such as the steepest descent and conjugate gradient methods; here we describe the latter. The coordinates $\mathbf{x}_n$ and $\mathbf{x}_{n+1}$ before and after the nth iteration are related [129] by

$$\mathbf{x}_{n+1} = \mathbf{x}_n + \mathbf{h}(\mathbf{x}_n)t \tag{5.16}$$

with

$$\mathbf{h}(\mathbf{x}_n) = -\mathbf{g}(\mathbf{x}_n) + \gamma_n \mathbf{h}(\mathbf{x}_{n-1})$$

where $\mathbf{g}(\mathbf{x}) = (\partial\varepsilon(\mathbf{x})/\partial x_1, \ldots, \partial\varepsilon(\mathbf{x})/\partial x_{2K}) = (g_1(\mathbf{x}), \ldots, g_{2K}(\mathbf{x}))$ is the gradient of the function $\varepsilon(\mathbf{x})$, t is the step of the gradient method, and

$$\gamma_n = -\frac{(\mathbf{g}(\mathbf{x}_n), \mathbf{g}(\mathbf{x}_n) - \mathbf{g}(\mathbf{x}_{n-1}))}{|\mathbf{g}(\mathbf{x}_{n-1})|^2}$$

According to Eqs (5.13)–(5.15) and the relation of symmetry, the components of the gradient vector at $\varphi = \pi$ can be represented in the form

$$\begin{aligned} g_i(\mathbf{x}) = 16\frac{(-1)^i}{\pi}\sum_{p=1}^{N}\frac{1}{p} \quad & (S_p\cos(2\pi p x_i) - C_p\sin(2\pi p x_i)) \\ & \times [I_p(\mathbf{x}) - \hat{I}_p] + 4(-1)^i(2Q-1)[I_0(\mathbf{x}) - \hat{I}_0] \end{aligned} \tag{5.17}$$

where S_p, C_p and Q are defined according to Eq. (5.15).

To define the step t of the gradient method, we consider the merit function along the conjugate direction as a function of t

$$\varepsilon[\mathbf{x}_n + \mathbf{h}(\mathbf{x}_n)t] = \varepsilon_1(t)$$

The optimum step size t can be determined by forming a second-order Taylor series expansion of $\varepsilon_1(t)$ about the point $t = 0$

$$\varepsilon_1(t) = \varepsilon_1(0) + \varepsilon_1'(0)t + \frac{\varepsilon_1''(0)t^2}{2}$$

The second-order expansion of $\varepsilon_1(t)$ takes a minimum at the t value given by

$$t = -\frac{\varepsilon_1'(0)}{\varepsilon_1''(0)} \tag{5.18}$$

Analytical formulae for the derivatives $\varepsilon_1'(0)$ and $\varepsilon_1''(0)$ can easily be obtained from Eqs (5.13)–(5.15).

The phase φ of the grooves can be treated as an additional degree of freedom in the optimization algorithm. According to Eqs (5.14) and (5.15), $\varphi = \pi$ is the optimal value with which to calculate the grating without the zero order. In all other cases, the needed value of the zero order is given by an appropriate choice of φ. To have equal intensity in the orders, let us define the value of φ in Eq. (5.14) as [59,127]

$$\varphi = 2\arcsin\left[\sqrt{\left(\frac{2N}{E_0 + 8Q(1-Q)N}\right)}\right] \tag{5.19}$$

where

$$E_0 = \sum_{\substack{j=-N \\ j\neq 0}}^{N} I_j$$

is the energy efficiency of the initial grating without zero order, i.e. the fraction of incident power appearing in the required orders. Then, from Eq. (5.14), it follows that the value of the zero order equals the mean value of the intensity in the orders. Taking into account the zero order, the energy efficiency increases by

$$\Delta E = E_0 \frac{2N + 1 - E_0 - \gamma N}{E_0 + \gamma N} \tag{5.20}$$

where $\gamma = 8Q(1 - Q)$.

5.1.3 Results of Numerical Computations

In Table 5.1 we present the results for binary gratings calculated for different numbers $M = 2N + 1$ of equal intensity diffraction orders. The r.m.s. error reported in Table 5.1 is defined as

$$\delta = \frac{1}{\bar{I}}\left[\frac{1}{M}\sum_{j=-N}^{N}(I_j - \bar{I})^2\right]^{1/2}$$

where

$$\bar{I} = \frac{1}{M}\sum_{j=-N}^{N} I_j$$

Table 5.1 Results of calculation for binary diffraction gratings

Number of orders M	Number of grooves K	Energy efficiency E (%)	r.m.s. error δ (%)
5	2	77.5	0.7
7	2	81.2	1.4
9	3	78.2	2.6
11	3	84.6	0.5
15	4	84.6	0.5
19	5	82.4	1.1
25	7	80.3	3.8
31	8	83.4	3.1
41	11	82.0	4.7
51	14	82.3	3.7
61	16	80.7	2.1
71	19	80.4	1.3
81	22	82.8	3.7
91	25	81.0	2.5
101	27	81.4	2.4
151	40	82.8	4.9
181	44	82.7	3.9
201	54	81.4	4.6
251	69	82.2	4.7
281	74	82.4	4.0

is a mean value. The high energy efficiency E and the low r.m.s. δ obtained for any grating of order up to 281 confirm the validity of the approach adopted. It should be noted that we may unrestrictedly use the method to compute gratings with 1001 orders or more still remaining within a few minutes of computation. We limit the presentation of the results to the gratings we have manufactured. Gratings with the number of orders $M > 15$ were calculated under the assumption that $\varphi = \pi$. During the calculation we have found that for $M \leq 15$ it is more convenient to form the zero order by choosing the value of φ according to Eq. (5.19). That choice of φ allows a 2–4% energy efficiency increase for gratings with 5 to 11 diffraction orders. To define the step t of the gradient method in each iteration, we used a quadratic approximation of the merit function along the conjugate directions. Then the value of t is the minimum of the approximating function. This choice greatly increased the speed of convergence.

For the asymmetrical gratings under study we noticed that the minimum groove width Δ_{min} of the initial binary profile defined in Eq. (5.11) remains approximately of the same size after the iteration correction. For example, for a 101-order grating we obtained a ratio Δ_{min}/period of 0.01.

Figure 5.3 shows the calculated profiles for a 101-order grating generating

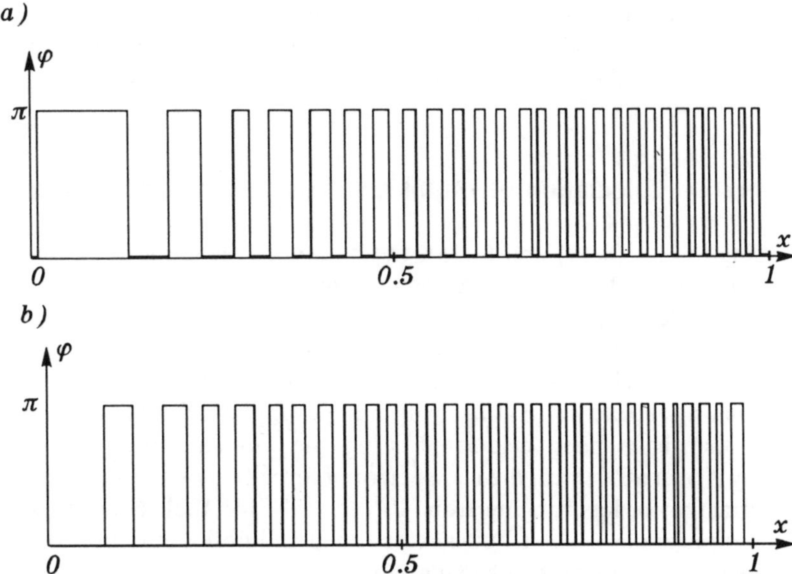

Figure 5.3 Calculated profile for a 101-order grating generating: (a) uniform intensity distribution; (b) linear intensity distribution

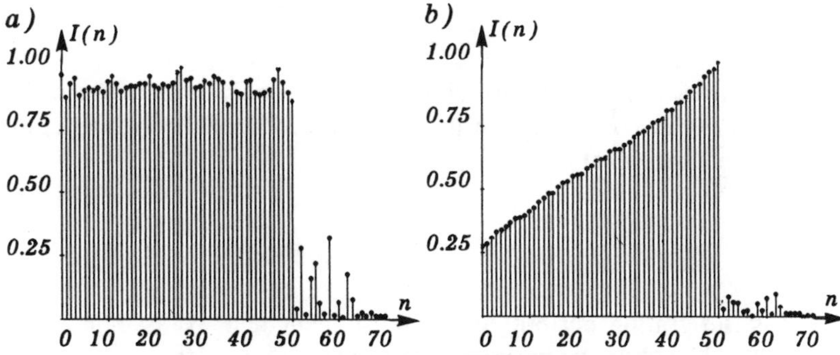

Figure 5.4 (a) Uniform intensity distribution for a 101-order grating; (b) linear intensity distribution for a 101-order grating

the uniform and linear intensity distributions shown in Fig. 5.4. Due to the symmetry, only half the diffraction orders are plotted.

We noticed that the initial approximation we are proposing gives a starting grating profile with the fill factor near 0.5; in any case, the iteration correction does not change the fill factor in a significant way. In the optimization algorithm we constrain the fill factor to ±10% of 0.5; this allows a minimum feature size down to 2λ with an acceptable scalar accuracy (close to 5%).

The proposed method can be generalized to the 2D case; here the initial approximation is a nonlinear binary predistortion of Eq. (5.6) applied to the phase function of the 2D DOE focusing into a rectangular area [130].

5.2 Iterative Calculation of Multilevel Phase Diffraction Gratings

According to Eqs (5.14) and (5.15), the binary diffraction gratings make it possible to form only symmetrical intensity distributions between diffraction orders. Multilevel phase diffraction gratings can produce an intensity distribution in diffraction orders which is non-symmetrical and as complex as one likes. The Gerchberg–Saxton (GS) algorithm is most commonly used for calculating diffraction gratings with multilevel phase function φ. For the gratings that have been calculated using the GS algorithm the energy efficiency is over 90%, with the r.m.s. error of formation of a desired intensity $\hat{I}_j$ in diffraction orders ranging from 10% to 15% [42,122]. Note, however, that for applications such as optical computing and fibre optics networks the error δ should be an order of magnitude smaller.

It has been shown in [20] that the GS algorithm can be treated as a gradient technique for minimizing the functional $\varepsilon(\varphi)$ representing the squared error

$$\varepsilon(\varphi) = \sum_{j=-\infty}^{\infty} (\sqrt{I_j} - \sqrt{\hat{I}_j})^2 \tag{5.21}$$

where the diffraction order intensities I_j are proportional to the squared modules of Fourier coefficients in the decomposition of the grating complex transmittance $\exp(i\varphi(x))$

$$I_j = \left| \frac{1}{d} \int_0^d \exp\left(i\varphi(x) - i\,\frac{2\pi}{d} jx \right) \mathrm{d}x \right|^2 \tag{5.22}$$

where d is the grating period. In the subsequent text we assume $d = 1$.

The GS algorithm features a stagnation effect: after several initial iterations the squared error's decrease becomes very slow. Both the type of the iterative procedure and its convergence depend on the choice of the error functional.

The rate of the iterative process convergence can be increased by changing the form of the functional in Eq. (5.21). Let us consider the gradient method of calculating the grating phase function $\varphi(x)$ for the error functional of an arbitrary form. Let $\varepsilon(\varphi)$ be a certain functional characterizing the distinction between the real intensity distribution in diffraction orders and a desired one

$$\varepsilon(\varphi) = \hat{\varepsilon}(\mathbf{I}, \mathbf{I}_d) \tag{5.23}$$

where $\mathbf{I}$ and $\mathbf{I}_d$ are the vectors whose components correspond to the calculated and required intensities in the diffraction orders. According to Eq. (5.22), the $\mathbf{I}$ vector components represent functionals of the grating phase function $\varphi(x)$. The challenge is to find a function $\hat{\varphi}(x)$ that can minimize the functional $\varepsilon(\varphi)$. The gradient method for minimizing the functional $\varepsilon(\varphi)$ suggests that the sought function $\hat{\varphi}(x)$ should be derived as a limit of the sequence of the functions $\varphi_n(x)$ specified in the form

$$\varphi_{n+1}(x) = \varphi_n(x) - t_n \varepsilon'(\varphi_n) \tag{5.24}$$

where $\varepsilon'(\varphi_n)$ is the functional gradient in the nth step of the iterative process of optimization and t_n is the step of the gradient method. Let us obtain the functional gradient $\varepsilon'(\varphi)$. To this end, consider an increment of the $\varepsilon(\varphi)$ functional caused by a small change $\Delta\varphi$ of phase. According to Eq. (5.23), the functional increment is given by

$$\Delta\varepsilon(\varphi) = \varepsilon(\varphi + \Delta\varphi) - \varepsilon(\varphi) = \sum_j \frac{\partial\hat{\varepsilon}(\mathbf{I}, \mathbf{I}_d)}{\partial I_j} \Delta I_j(\varphi) \tag{5.25}$$

The following relation holds correct up to second-order terms

$$\Delta I_j(\varphi) = 2\,\mathrm{Re}(F_j^*(\varphi)\Delta F_j(\varphi)) \tag{5.26}$$

where * denotes complex conjugation and

$$F_j(\varphi) = \int_0^1 \exp(i\varphi(x) - i2\pi jx)\,\mathrm{d}x \tag{5.27}$$

is the complex amplitude of field in the jth diffraction order.

By virtue of Eq. (5.27), $\Delta F_j(\varphi)$ takes the form

$$\begin{aligned}\Delta F_j(\varphi) &= \int_0^1 \{\exp[i\varphi(x) + i\Delta\varphi(x)] - \exp[i\varphi(x)]\}\exp(-i2\pi jx)\,\mathrm{d}x \\ &\approx \int_0^1 i\Delta\varphi(x)\exp[i\varphi(x) - i2\pi jx]\,\mathrm{d}x\end{aligned} \tag{5.28}$$

Substituting Eqs (5.26)–(5.28) in Eq. (5.25), we obtain the functional increment in the form

$$\Delta\varepsilon(\varphi) = 2\,\mathrm{Re}\left(\sum_j \frac{\partial\hat{\varepsilon}(\mathbf{I}, \mathbf{I}_d)}{\partial I_j} F_j^*(\varphi) \int_0^1 i\Delta\varphi(x)\exp(i\varphi(x) - i2\pi jx)\,\mathrm{d}x\right) \tag{5.29}$$

According to Eq. (5.29), the functional gradient $\varepsilon'(\varphi)$ is specified in the form

$$\varepsilon'(\varphi) = 2\,\mathrm{Re}\left(i \sum_j \frac{\partial\hat{\varepsilon}(\mathbf{I}, \mathbf{I}_d)}{\partial I_j} F_j^*(\varphi)\exp(i\varphi(x) - i2\pi jx)\right) \tag{5.30}$$

For convenience of subsequent analysis, Eq. (5.30) may be rearranged to give

$$\varepsilon'(\varphi) = 2\,\mathrm{Re}\{i\exp[i\varphi(x)]\Psi^*(x)\} = -2|\Psi(x)|\sin\{\varphi(x) - \arg[\Psi(x)]\} \quad (5.31)$$

where

$$\Psi(x) = \int_{-\infty}^{\infty} f(\xi)\exp(i2\pi x\xi)\,\mathrm{d}\xi$$

is the inverse Fourier transform of the function

$$f(\xi) = \sum_j \frac{\partial\hat{\varepsilon}(\mathbf{I}, \mathbf{I}_d)}{\partial I_j} F_j(\varphi)\,\delta(\xi - j) \quad (5.32)$$

where $\delta(\xi)$ is the delta function.

The calculation of the $\Psi(x)$ function bears a resemblance to the three initial steps of the Gerchberg–Saxton algorithm. Actually, the calculation of $\Psi(x)$ in Eq. (5.31) reduces to the implementation of the following steps: (1) the calculation of the Fourier transform aimed at obtaining the values F_j of the field in diffraction orders, (2) replacement of F_j by the function $f(\xi)$, (3) implementation of the inverse Fourier transform of the function $f(\xi)$. In the GS algorithm, the F_j values of the field are replaced not by the $f(\xi)$ function but by the values

$$\hat{F}_j = \sqrt{\hat{I}_j} \cdot F_j/|F_j|$$

To define the step t of the gradient method, we consider the functional $\varepsilon(\varphi)$ along the antigradient direction as a function of t

$$\varepsilon(\varphi) = \varepsilon[\varphi - t\varepsilon'(\varphi)] = \varepsilon_1(t)$$

The step t optimum size can be determined by expanding the function $\varepsilon_1(t)$ into a first-order Taylor series about the point $t = 0$

$$\varepsilon_1(t) = \varepsilon_1(0) + \varepsilon_1'(0)t \quad (5.33)$$

The first-order expansion of $\varepsilon_1(t)$ is equal to zero at t_f, given by

$$t_f = -\frac{\varepsilon_1(0)}{\varepsilon_1'(0)} \quad (5.34)$$

The derivative $\varepsilon_1'(0)$ can be derived from Eqs (5.29) and (5.30) in the form

$$\varepsilon_1'(0) = -4\int_0^1 |\Psi(x)|^2 \sin^2\{\varphi(x) - \arg[\Psi(x)]\}\,\mathrm{d}x \quad (5.35)$$

If we assume the function $\varepsilon_1(t)$ to be quadratic along the antigradient direction, it is easy to obtain the estimate for the optimal step t_s, in the form

$$t_s = -\frac{[\varepsilon_1(0)]^2}{2\varepsilon_1(t_f)\,\varepsilon_1'(0)} \quad (5.36)$$

To find t_s, we will need to recalculate order intensities for the calculation of $\varepsilon_1(t_f)$. When the convergence rate is stabilized, the functional $\varepsilon(\varphi)$ shows weak variations from iteration to iteration. In this case $\varepsilon_1(t_f) \approx \varepsilon_1(0)$, and we can write

$$t_s = \frac{t_f}{2} = \varepsilon_1(0)\left(8\int_0^1 |\Psi(x)|^2 \sin^2\{\varphi(x) - \arg[\Psi(x)]\}\,\mathrm{d}x\right)^{-1} \quad (5.37)$$

It is noteworthy that the resulting relations for the gradient and the step size are the same for any functional $\varepsilon(\varphi)$.

Let us consider the interrelation between the GS algorithm and the gradient procedure of minimizing the functional of Eq. (5.21). The minimization of this functional appears to be equivalent to the maximization of the functional given by

$$\bar{\varepsilon}(\varphi) = 2\sum_j \sqrt{(I_j \hat{I}_j)} \quad (5.38)$$

The gradient of the functional in Eq. (5.38) is decided by Eq. (5.31) for the $f(\xi)$ function taking the form

$$f(\xi) = \sum_j \sqrt{(\hat{I}_j)}\,\frac{F_j(\varphi)}{|F_j(\varphi)|}\,\delta(\xi - j) \quad (5.39)$$

where the function $f(\xi)$ corresponds to the replacement of Eq. (1.6) conducted in a Fourier plane for the GS algorithm. Accordingly, the calculating procedure for the function $\Psi(x)$ in Eq. (5.31) coincides with the first three steps of the GS algorithm. A distinction of the gradient procedure specified by Eqs (5.24), (5.31) and (5.39) from the GS algorithm is the calculation of a new phase approximation $\varphi_{n+1}(x)$ in the nth step of the iteration procedure. For the GS algorithm we have $\varphi_{n+1}(x) = \arg[\Psi_n(x)]$, whereas for the gradient procedure $\varphi_{n+1}(x)$ can be derived from the relation in Eq. (5.24). Now consider a situation when several iterations have been performed and the error $\varepsilon(\varphi)$ is changing slowly. In that case, for most values of x we have

$$\sin\{\varphi_n(x) - \arg[\Psi_n(x)]\} \approx \varphi_n(x) - \arg[\Psi_n(x)] \quad (5.40)$$

From Eqs (5.24) and (5.31), it is seen that for the gradient method the difference $\varphi_{n+1}(x) - \varphi_n(x)$ is approximately proportional to $\varphi_n(x) - \arg[\Psi_n(x)]$. Considering that $\arg[\Psi_n(x)]$ is the new phase that would be given by the GS algorithm, one can conclude that the gradient procedure of Eqs (5.24), (5.31), and (5.39) gives a new phase that is going in approximately the same direction as a GS algorithm [20]. For the gradient procedure, the gradient of Eqs (5.31) and (5.39) differs somewhat from $\varphi_n(x) - \arg[\Psi_n(x)]$ due to the weighting factor $|\Psi_n(x)|$.

As distinct from the GS algorithm, the gradient procedure of Eqs (5.24) and (5.31) permits the choice of various functionals $\varepsilon(\varphi)$.

The form of the functional $\varepsilon(\varphi)$ has an essential effect on the iterative process convergence. It is appropriate to investigate the convergence of the gradient procedure for the error functional of the form

$$\varepsilon(\varphi, p) = \sum_j (I_j - \hat{I}_j)^p \tag{5.41}$$

In comparison with the functional of Eq. (5.21), the functional $\varepsilon(\varphi, p)$ of Eq. (5.41), for $p > 1$, shows higher sensitivity to the deviation of intensity distribution in diffraction orders from the required distribution. The gradient of the functional of Eq. (5.41) is defined by Eq. (5.31), with the function $f(\xi)$ having the form

$$f(\xi) = p \sum_j (I_j - \hat{I}_j)^{p-1} F_j \delta(\xi - j) \tag{5.42}$$

To choose appropriately an optimal technique for calculating multiorder diffraction gratings, we shall analyze the robustness of the following iterative procedures: the GS algorithm, the adaptive–additive (AA) algorithm dealt with in Chapter 1, and the gradient method specified by Eqs (5.24) and (5.31) for the functional of Eq. (5.41).

Let us recall that the only difference between the GS and the AA algorithms is in the form of the replacement of calculated values of the field complex amplitude F_j in diffraction orders. In the GS algorithm, the values of F_j are replaced by the values $\hat{F}_j$ according to the rule

$$\hat{F}_j = \begin{cases} \sqrt{(\hat{I}_j)} \dfrac{F_j}{|F_j|}, & j \in Q \\ 0, & \text{else} \end{cases} \tag{5.43}$$

where $Q = \{n_1, n_2, \ldots, n_M\}$ are the indices of the required diffraction orders. In the AA algorithm, the replacement of Eq. (5.43) takes the form

$$\hat{F}_j = \begin{cases} |\alpha \sqrt{(\hat{I}_j)} + (1-\alpha)\sqrt{(I_j)}| \dfrac{F_j}{|F_j|}, & j \in Q \\ F_j, & \text{else} \end{cases} \tag{5.44}$$

where α is a coefficient to be chosen.

To characterize the grating operation, we shall employ the value of the energy efficiency E

$$E = \sum_{j \in Q} I_j$$

and the r.m.s. error δ

$$\delta = \frac{1}{\bar{I}} \left[\frac{1}{M} \sum_{j \in Q} (I_j - \hat{I}_j)^2 \right]^{1/2}$$

Table 5.2 Results of iterative calculation of a multiorder diffraction grating

Number of orders	GS algorithm		AA algorithm		Gradient method for functional of Eq. (5.41) at $p = 2$	
	E (%)	δ (%)	E (%)	δ (%)	E (%)	δ (%)
5	90.1	14.7	88.5	0.1	91.5	0.5
9	95.6	17.5	93.5	0.2	93.8	0.2
15	97.9	15.5	93.6	0.1	94.0	0.8
19	97.2	14.9	94.1	0.1	94.5	0.7
23	95.1	16.9	93.9	0.1	94.8	0.6
27	98.3	18.7	93.3	0.1	94.4	1.2
33	97.9	14.8	94.5	0.1	96.2	1.8
39	99.2	11.5	94.2	0.2	96.3	1.5
45	99.6	10.8	96.1	0.4	97.5	1.2
51	99.7	9.8	95.9	0.7	98.3	1.8

where

$$\bar{I} = \frac{1}{M}\sum_{j\in Q} I_j$$

is an average value and M is the number of orders.

Table 5.2 gives calculated results for diffraction gratings with different numbers of symmetrically located orders of equal intensity. The calculation was carried out using the fast Fourier transform, the number of pixels of grating phase function being $n = 64$. To take account of diffraction by the raster, the intensities $\hat{I}_j$ were assumed to be not uniform but in inverse proportion to the function $f(j) = \text{sinc}^2(\pi j/n), j = -n/2, n/2$ [42]. In the course of calculation, 80 to 100 iterations proved sufficient to stabilize errors in the iteration procedures. As the starting approximation, we used a stochastic phase uniformly distributed in the interval $[0, 2\pi]$.

Table 5.2 gives values of E and δ for the gratings calculated via the GS algorithm (the replacement of Eq. (5.43)) and via the AA algorithm (the replacement of Eq. (5.44)). Note that in this case the first 15–20 iterations (as long as the deviation of the calculated intensity in orders from the desired one decreases rapidly) were conducted using the GS algorithm with the replacement in Eq. (5.43). As soon as the error in the GS algorithm had been stabilized, we changed to the AA algorithm, Eq. (5.44), for $\alpha = 2$. The table also gives values of E and δ for the gratings calculated using the gradient method specified by Eqs (5.24), (5.31) and (5.32) for the functional of Eq. (5.41) at $p = 2$. The optimum step size t_n in Eq. (5.24) was determined according to Eqs (5.36) and (5.37). In the course of the first 10–20 iterations, the step t_n should be derived from Eq. (5.36), because the use of a simplified

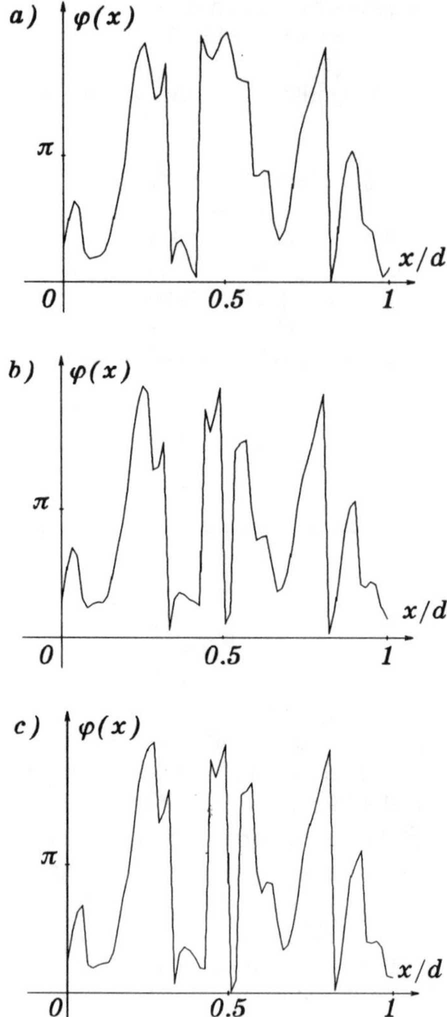

Figure 5.5 Phase functions of 11-order gratings derived via (a) the GS algorithm; (b) the AA algorithm; (c) the gradient method

formula (5.37) for calculating the step t_n leads to the oscillation of the criteria in Eq. (5.41). The calculation of t_n based on Eq. (5.36) requires taking an extra Fourier transform. To simplify the calculational procedure, the first 10–20 iterations should be performed using the GS algorithm; in later steps, the use of the gradient procedure results in smooth variations of the functional $\varepsilon(\varphi, 2)$ so that the use of the simplified formula (5.37) for calculating the step t_n provides the stable convergence of the iteration procedure. From the data given in Table 5.2, it is seen that the gratings

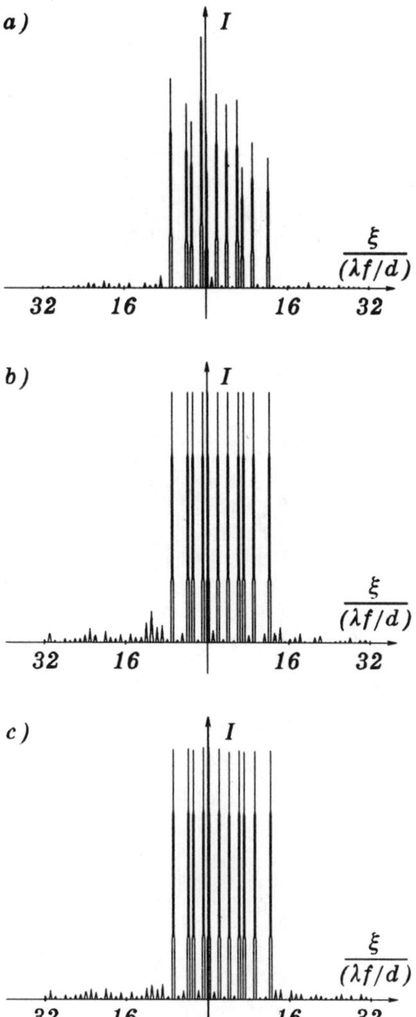

Figure 5.6 Intensity distribution in the Fourier plane for 11-order gratings consisting of four periods: (a) the GS algorithm; (b) the AA algorithm; (c) the gradient method

computed via the GS algorithm possess the r.m.s. error $\delta = 10$–15% with a 90% to 99% energy efficiency. Note that the use of the gradient method for the functional of Eq. (5.21) did not result in an essentially improved convergence, whereas the use of the AA algorithm allows one to reduce the error δ to 0.1–0.2%, with the energy efficiency E to be reduced only by 1–3%.

For the gratings calculated via the gradient method for the functional

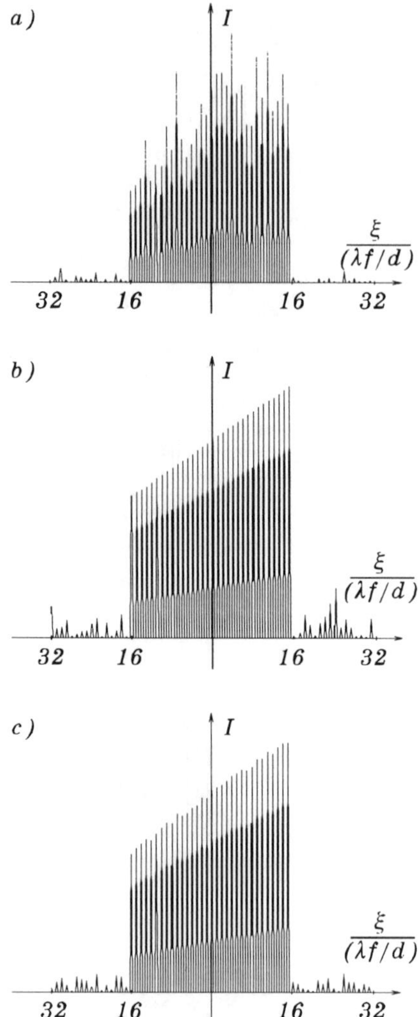

Figure 5.7 Linear intensity distribution in the Fourier plane for 32-order gratings calculated using (a) the GS algorithm; (b) the AA algorithm; (c) the gradient method

$\varepsilon(\varphi, 2)$, the error δ is nearly an order of magnitude smaller than for the gratings calculated via the GS algorithm. From numerical calculations, the convergence of the gradient method for the functional $\varepsilon(\varphi, p)$ is seen to be the best for $p = 2$. With increasing p the rate of convergence decreases.

The calculation has also been carried out for gratings with non-symmetrical location of orders. Figure 5.5 depicts phase functions of the diffraction gratings producing 11 non-symmetrically located orders of equal intensity.

The phase function shown in Fig. 5.5a has been calculated using the GS algorithm, that in Fig. 5.5b has been derived via the AA algorithm, and that in Fig. 5.5c has been obtained using the gradient method for minimizing the functional $\varepsilon(\varphi, 2)$. Intensity distributions in the Fourier plane for the gratings shown in Fig. 5.5 and consisting of four periods are illustrated in Fig. 5.6. Values of δ and E are, respectively: 20.4% and 94.3% (Fig. 5.6a), 0.1% and 90.2% (Fig. 5.6b), and 0.6% and 91.2% (Fig. 5.6c).

Another example of asymmetry is presented in Fig. 5.7. Intensity distributions in the Fourier plane shown in Fig. 5.7 have been obtained for 32-order gratings aimed at generating a linear intensity distribution in orders. Values of δ and E are, respectively: 17.1% and 98.1% for the grating calculated using the GS algorithm (Fig. 5.7a); 0.3% and 94.1% for the grating calculated using the AA algorithm (Fig. 5.7b); and 1.4% and 95.3% for the grating calculated using the gradient method for the functional $\varepsilon(\varphi, 2)$ (Fig. 5.7c).

Calculational results (Table 5.2 and Figs 5.6 and 5.7) testify that for the purpose of calculating multiorder diffraction gratings, the AA algorithm and the gradient method for the $\varepsilon(\varphi, 2)$ functional are essentially superior to the GS algorithm. The rate of convergence of the adaptive procedure of Eq. (5.44) is somewhat higher than that of the gradient method. The gradient method is also found to be more complex in terms of software realization, because it requires the explicit calculation of the gradient and the step size in each iteration.

6

Iterative Methods for Calculating Multifocus DOEs

A multiorder diffraction grating splits a plane light beam into a set of outgoing plane beams with predetermined energy distributions. Each beam is defined by a different order in the grating equation $j\lambda = d \sin\theta$, where j is the order number, λ is the wavelength, d is the grating period, and θ is the angle at which the departing beams leave the grating. Multiorder DOEs which, due to diffraction, generate a set of outgoing beams (orders) with different profiles of phase and with different energy distributions can be considered as a generalization of the multiorder grating.

The phase function of a single phase DOE can have the property of a combination of two or more single optical elements. One possible approach to the problem of designing multiorder DOEs is to combine a lens and a multiorder diffraction grating [114,116,131]. In the focal plane of the lens a set of points with a predetermined power distribution is given [114,116]. More complicated multiorder DOEs such as a combination of focusators [37,126,132,133] (thin phase optical elements focusing into a line of desired shape) and diffraction gratings ensure the formation and replication of nonpoint curve-like images in a single plane.

In this chapter we consider the computation of multiorder DOEs aimed at focusing into a set of lines that have pronounced multifocal properties. Multifocal properties imply that the focusing occurs in different planes. Such multifocus DOEs play an important part in such applications as image processing, optical computing, 3D image formation systems, intraocular lenses with accommodation abilities, and laser optical disk heads.

In section 6.1 we deal with multifocus focusators – phase optical elements focusing into a set of scaled lines with desired power ratios and located in different focal planes along the optical axis. As a special case, we also consider the multiorder focusators aimed at focusing into a set of scaled lines in a single focal plane. Multifocus zone plates and combined multifocus elements intended to focus in two sets of lines simultaneously are considered in section 6.2. Section 6.3 is dedicated to computing Fresnel-type multifocus diffractive lenses. In section 6.4 we consider two-order DOEs focusing into two different focal lines in the two main diffraction orders.

6.1 Multifocus Focusators

6.1.1 *Ray-tracing Solution to a Problem of Focusing into a Line*

Computer-aided generation of phase diffractive elements offers an opportunity to generate an arbitrary phase retardation. Zoned phase microrelief with limited height performs a phase retardation equivalent to a wide-range-varying phase function φ of a thin optical element. In this case, the microrelief height is proportional to the phase function φ modulo 2π.

A method for computing multifocus diffractive elements presented in this chapter uses the theory of focusators. Therefore, we first briefly consider focusators focusing into a line. Let a line L in the focal plane (x, y) be given by the parametric equation

$$\begin{cases} x = x(\xi) \\ y = y(\xi) \end{cases} \tag{6.1}$$

where $\xi \in [0, S]$ is the natural parameter (see Fig. 6.1).

If the incident beam has an intensity $I_0(\mathbf{u})$ and a constant phase, we can apply the ray-tracing method to obtain the phase function of the focusator, in paraxial approximation, in the form [37,126,132,133]

$$\varphi(\mathbf{u}) = -\frac{k\mathbf{u}^2}{2f} + \hat{\varphi}(\mathbf{u}) \tag{6.2}$$

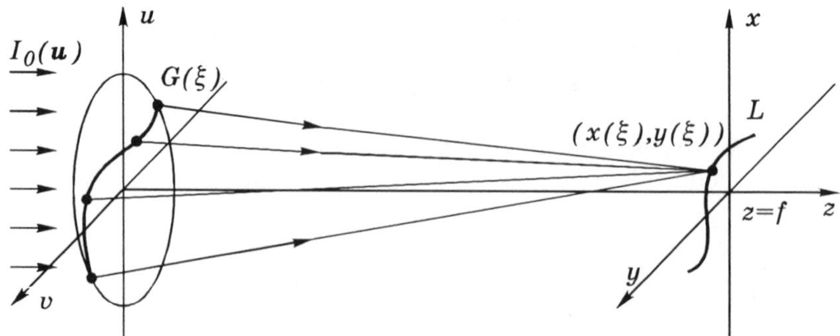

Figure 6.1 Calculation of a focusator focusing into a line

where

$$\hat{\varphi}(\mathbf{u}) = \frac{k}{f}\int_{u_0}^{u} x[\xi(u, v_0)]\,\mathrm{d}u + \frac{k}{f}\int_{v_0}^{v} y[\xi(u, v)]\,\mathrm{d}v \tag{6.3}$$

$\mathbf{u} = (u, v)$ represents the coordinates in the plane of the focusator, $k = 2\pi/\lambda$, λ is the wavelength, and f is the focusator-to-focal plane distance. The function $\xi(\mathbf{u})$ in Eq. (6.3) describes smooth correspondence between the plane of the focusator and the points on the focal line L. The expression of $\xi(\mathbf{u})$ depends on the illuminating beam intensity and on the required energy distribution $\theta(\xi)$, $\xi \in [0, S]$, along the line L. As is shown in [37,126,132], the rays focused in each point $(x(\xi), y(\xi))$ of the focal line L are traced from the smooth curve $G(\xi)$ on the focusator's aperture called the layer (see Fig. 6.1). The specific form of the $\xi(\mathbf{u})$ function is sought from the layer equation which (in paraxial approximation) is given by [37,126]

$$u\frac{\mathrm{d}x(\xi)}{\mathrm{d}\xi} + v\frac{\mathrm{d}y(\xi)}{\mathrm{d}\xi} = d(\xi) \tag{6.4}$$

The $d(\xi)$ function in Eq. (6.4) is specified by a required energy distribution $\theta(\xi)$, $\xi \in [0, S]$, along the focal line L. By using the law of light flux conservation, one can find the equation for the $\mathrm{d}(\xi)$ function in the form

$$\iint_{(G(0),G(\xi))} I_0(\mathbf{u})\,\mathrm{d}^2\mathbf{u} = \int_0^{\xi} \theta(\zeta)\,\mathrm{d}\zeta \tag{6.5}$$

where $(G(0), G(\xi))$ denotes the part of the focusator's aperture lying between the layer $G(0)$ corresponding to the initial point of line L and the current layer $G(\xi)$.

Focusing into a straight-line segment represents the most important practical realization of focusing into a line. Therefore, we have found it reasonable to enlarge on the calculational equations for the phase function of a focusator into a straight-line segment. Let us assume that the focusator's aperture G is bounded by the curves $v = g_1(u)$ and $v = g_2(u)$, and by the straight-line segment $u = a$ and $u = b$. Let $\theta(x)$ denote a desired energy distribution along the focal segment $|x| \le d$. In that case, manipulation of Eqs (6.2) to (6.5) easily yields the phase function of the focusator into a straight-line segment in the form [134–136]

$$\varphi(\mathbf{u}) = -\frac{k\mathbf{u}^2}{2f} + \frac{k}{f}\int_0^{u} \chi(\gamma)\,\mathrm{d}\gamma \tag{6.6}$$

where the $\chi(u)$ function can be found from the differential equation

$$\frac{\mathrm{d}\chi(u)}{\mathrm{d}u} = \frac{1}{\theta[\chi(u)]}\int_{g_1(u)}^{g_2(u)} I_0(u, v)\,\mathrm{d}v \tag{6.7}$$

with boundary conditions given by $\chi(a) = -d$, $\chi(b) = d$.

Now let us prove that the phase function of a focusator focusing into a line L_p with parametric equations

$$\begin{cases} x_p = px(\xi_p/p) \\ y_p = py(\xi_p/p) \end{cases} \tag{6.8}$$

where $\xi_p \in [0, pS]$ is a natural parameter, has the form

$$\varphi_p(\mathbf{u}) = -\frac{k\mathbf{u}^2}{2f} + p\hat{\varphi}(\mathbf{u}) \tag{6.9}$$

In addition, the energy distribution $\theta_p(\xi_p)$ along the line L_p is given by

$$\theta_p(\xi_p) = \frac{1}{p}\theta(\xi_p/p), \quad \xi_p \in [0, pS] \tag{6.10}$$

which means that it coincides with the $\theta(\xi)$ function up to scale.

Proof

Let $\xi_p(\mathbf{u})$ be a function of the ray correspondence for the focusator focusing into a line L_p. The layer equation (6.4) for the focal line of Eq. (6.8) takes the form

$$u\frac{\mathrm{d}x(\xi_p/p)}{\mathrm{d}(\xi_p/p)} + v\frac{\mathrm{d}y(\xi_p/p)}{\mathrm{d}(\xi_p/p)} = d_p(\xi_p) \tag{6.11}$$

Based on direct substitution, it is easy to show that Eq. (6.11) turns into the identity for the functions $d_p(\xi_p), \xi_p(\mathbf{u})$ given by

$$\begin{cases} \xi_p(\mathbf{u}) = p\xi(\mathbf{u}) \\ d_p(\xi_p) = d(\xi_p/p) \end{cases} \tag{6.12}$$

Let us determine the form of the function $\theta_p(\xi_p)$ for the functions $d_p(\xi_p)$ and $\xi_p(\mathbf{u})$ given in Eq. (6.12). According to Eqs (6.11) and (6.12), for the focusators focusing into the lines L and L_p the layers for the points $(x(\xi), y(\xi))$ and $(x_p(p\xi), y_p(p\xi))$ coincide. Then the law of light flux conservation gives

$$\int_0^{\xi} \theta(\zeta)\,\mathrm{d}\zeta = \int_0^{p\xi} \theta_p(\zeta)\,\mathrm{d}\zeta \tag{6.13}$$

From Eq. (6.13), we find that

$$\theta_p(\xi_p) = \frac{1}{p}\theta(\xi_p/p), \quad \xi_p \in [0, pS]$$

i.e. the function $\theta_p(\xi_p)$ coincides with the $\theta(\xi)$ function up to scale. Substituting Eqs (6.8) and (6.12) into the general relationships of Eqs (6.2) and (6.3), we obtain Eq. (6.9), as we wished to prove. Thus, the

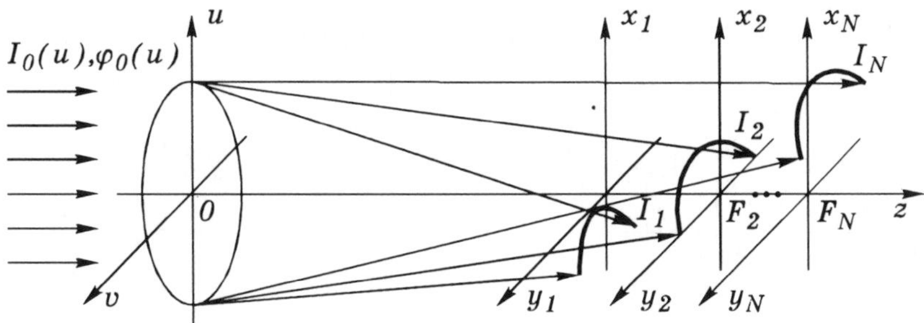

Figure 6.2 Geometry of the focusing process realized by the multifocus focusator

multiplication of the $\hat{\varphi}(\mathbf{u})$ function by the constant p results in focusing in the p times scaled line L. The above brief description of focusators focusing into a line appears to be sufficient for purposes of further consideration.

6.1.2 Multifocus Focusator Elaboration

The multifocus DOEs dealt with in this section are dedicated to focusing the illuminating beam into N scaled lines L located in different focal planes along the optical axis. Note in addition that between the lines there should be a predetermined power distribution $I_1, \ldots, I_N, (\Sigma_{i=1}^{N} I_i = 1)$ (see Fig. 6.2).

It is known that a phase nonlinearity in the phase function of the element yields additional diffraction orders [47,137–141]. The basic idea in designing multifocus DOEs is to use the diffraction orders generated by applying a nonlinear predistortion to the phase function given in the general form of Eqs (6.2) and (6.3). For this kind of phase, the diffraction patterns correspond to a set of scaled lines located in different focal planes. The kind of nonlinearity influences the energy distribution between the lines. Here, we propose a method for choosing the nonlinearity that ensures a required energy distribution between a preset number of lines (one line in each plane). With this in mind, we represent the phase function of a multifocus DOE in the form [139–142]

$$\varphi_{mf}(\mathbf{u}) = \varphi_1(\mathbf{u}) + \varphi_{11}(\mathbf{u}) + \Phi[\varphi_d(\mathbf{u})] - \varphi_0(\mathbf{u}), \quad \mathbf{u} \in G \tag{6.14}$$

where

$$\varphi_d(\mathbf{u}) = \mathrm{mod}_{2\pi}[\varphi_2(\mathbf{u}) + \varphi_{22}(\mathbf{u}) + \varphi_{23}(\mathbf{u})] \tag{6.15}$$

Here, $\varphi_0(\mathbf{u})$ is the phase of a beam illuminating the DOE with aperture G, the functions $\varphi_1(\mathbf{u})$ and $\varphi_2(\mathbf{u})$, defined according to Eqs (6.2) and (6.3),

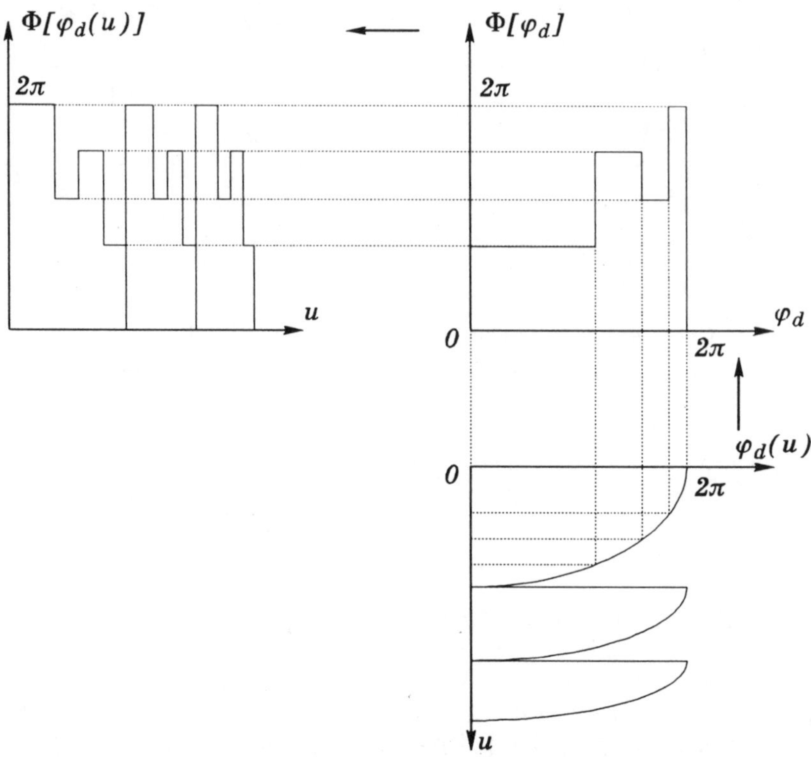

Figure 6.3 Nonlinear phase predistortion (1D case)

are the paraxial phases of the focusators focusing into the line L in some focal planes $z = f_1$ and $z = f_2$, and the functions

$$\varphi_{11}(\mathbf{u}) = \frac{k}{f_1}\mathbf{x}_1\mathbf{u}, \quad \varphi_{22}(\mathbf{u}) = \frac{k}{f_2}\mathbf{x}_2\mathbf{u}, \quad \varphi_{23}(\mathbf{u}) = -\frac{k}{2f_3}\mathbf{u}^2 \tag{6.16}$$

represent the paraxial phases of the prisms and the lenses, respectively.

The term $\varphi_0(\mathbf{u})$ compensates for the phase of the illuminating beam. The focusator phase $\varphi_1(\mathbf{u})$, the prism phases $\varphi_{11}(\mathbf{u})$ and $\varphi_{22}(\mathbf{u})$, and the lens phase $\varphi_{23}(\mathbf{u})$ have been included to allow for the additional degrees of freedom to choose the coordinates of focal planes, line locations in these planes, and a scaling factor. The function $\Phi[\varphi_d(\mathbf{u})] \in [0, 2\pi]$ describes a nonlinear predistortion of the phase $\varphi_d(\mathbf{u})$ modulo 2π. The mechanism of nonlinear predistortion is shown in Fig. 6.3. The function $\Phi[\varphi_d(\mathbf{u})]$ is then a kind of zone plate that can be binary or staircase or even have a continuously varying microrelief.

According to our design, we shall call the DOE with the phase of Eqs (6.14) and (6.15) the 'multifocus focusator'. To describe the operation of the multifocus focusator, we consider the case of an illuminating beam of the kind

$$W_0(\mathbf{u}) = \sqrt{[I_0(\mathbf{u})]}\,e^{i\varphi_0(\mathbf{u})}$$

Accordingly, the field immediately beyond the plane of the multifocus focusator becomes

$$W(\mathbf{u}) = \sqrt{[I_0(\mathbf{u})]}\exp\{i\varphi_1(\mathbf{u}) + i\varphi_{11}(\mathbf{u}) + i\Phi[\varphi_d(\mathbf{u})]\} \tag{6.17}$$

In order to describe the formation of a set of lines, we expand the function $\exp(i\Phi[\xi])$ into a Fourier series on the $[0, 2\pi]$ interval [139–142]

$$\exp(i\Phi[\xi]) = \sum_{j=-\infty}^{\infty} c_j \exp(ij\xi) \tag{6.18}$$

where

$$c_j = \frac{1}{2\pi}\int_0^{2\pi} \exp(i\Phi[\xi] - ij\xi)\,\mathrm{d}\xi \tag{6.19}$$

are the Fourier coefficients and $\Sigma_{j=-\infty}^{\infty}|c_j|^2 = 1$.

Substituting $\xi = \varphi_d(\mathbf{u})$ into Eq. (6.18) and using the 2π phase periodicity, Eq. (6.17) can be recast in the form

$$W(\mathbf{u}) = \sqrt{[I_0(\mathbf{u})]}\exp[i\varphi_1(\mathbf{u}) + i\varphi_{11}(\mathbf{u})] \times \sum_{j=-\infty}^{\infty} c_j \exp\{ij[\varphi_2(\mathbf{u}) + \varphi_{22}(\mathbf{u}) + \varphi_{23}(\mathbf{u})]\} \tag{6.20}$$

We can read Eq. (6.20) as a superposition of beams with the following phase functions

$$\varphi_j(\mathbf{u}) = -\frac{k\mathbf{u}^2}{2F_j} + p_j\left[\frac{k}{F_j}\int_{u_0}^{u} x[\xi(\mathbf{u})]\,\mathrm{d}u + \frac{k}{F_j}\int_{v_0}^{v} y[\xi(\mathbf{u})]\,\mathrm{d}v\right] + \frac{k}{F_j}\mathbf{X}_j\mathbf{u} \tag{6.21}$$

where

$$F_j = \frac{f_1 f_2}{f_2 + jf_1(1 + f_2/f_3)} \tag{6.22}$$

$$p_j = \frac{f_2 + jf_1}{f_2 + jf_1(1 + f_2/f_3)} \tag{6.23}$$

$$\mathbf{X}_j = \frac{\mathbf{x}_1 f_2 + j\mathbf{x}_2 f_1}{f_2 + jf_1(1 + f_2/f_3)} \tag{6.24}$$

The squared modules of the Fourier coefficients, Eq. (6.19), represent the power distribution between the beams, and the functions of Eq. (6.21) define the structure of the images formed in the diffraction orders of the multifocus focusator. According to the general relationships of Eqs (6.2) and (6.3), the phase function $\varphi_j(\mathbf{u})$ corresponds to focusing in a line L_{p_j} in the plane $z = F_j$. The line L_{p_j} is defined in Eq. (6.8) and is the line L scaled by factor p_j. The last term in Eq. (6.21) corresponds to a phase function of a prism that gives a shift $\mathbf{X}_j$ of the line L_{p_j} in the focal plane $z = F_j$.

It should be noted that the lines formed by the multifocus focusator are not completely independent. The images generated in the plane $z = F_j$ by the beams with phase functions $\varphi_k(\mathbf{u})$, for $k \neq j$, correspond to the defocused lines. In general, the width of defocused lines is one order of magnitude greater than the diffractive width of the line L_{p_j}, which makes their influence on the L_{p_j} line negligibly small. In some rare cases when the interaction cannot be neglected, an appropriate choice of $\mathbf{x}_1$ and $\mathbf{x}_2$ in Eq. (6.24) allows shifting of the defocused lines relative to useful ones, resulting in the independent formation of focal lines. Therefore, in the following, we neglect an interaction between the images formed in the orders of the multifocus focusator.

Supposing that $j_1, \ldots, j_N$ are the indexes corresponding to the required focal planes, it is sufficient to determine the function $\exp(i\Phi[\xi]), \xi \in [0, 2\pi]$, with the Fourier coefficients (see Eq. (6.19)) obeying the equations

$$|c_{j_i}|^2 = I_i, \quad i = 1, \ldots, N \tag{6.25}$$

We propose that the predistortion $\Phi[\xi]$ should be interpreted as the phase function of the N-order diffraction grating with a period of 2π and intensities $I_1, \ldots, I_N$ in the orders $j_1, \ldots, j_N$ [47,139–142]. This interpretation reduces the problem of computing the nonlinear predistortion $\Phi[\xi]$ to the well-known problem of synthesizing the phase diffraction grating with a preset power distribution between diffraction orders. The grating energy efficiency

$$E = \sum_{i=1}^{N} |c_{j_i}|^2$$

specifies which portion of the illuminating beam energy is focused into the required focal lines. The deviation of the intensity values in the orders $j_1, \ldots, j_N$ from the desired values $I_1, \ldots, I_N$ characterizes the error in the formation of the required energy distribution between focal lines. The calculational techniques considered in Chapter 5 make it possible to calculate multiorder diffraction gratings characterized by an 80–95% energy efficiency, with the r.m.s. deviation of order intensities from the desired ones being within several per cent.

Let us now consider some special cases of multifocus focusators. In the case of

$$f_3 = \infty(\varphi_{23}(\mathbf{u}) \equiv 0) \tag{6.26}$$

in Eqs (6.14) and (6.15), the coefficients p_j of Eq. (6.23) describing the scaling of focal lines are equal to 1. Accordingly, the sizes of the focal lines are the same in all focal planes.

In the case of

$$f_3 = -f_2 \tag{6.27}$$

in Eqs (6.14) and (6.15), the functions in Eq. (6.21) defining the structure of the images formed in the diffraction orders of the multifocus focusator take the form

$$\varphi_j(\mathbf{u}) = -\frac{k\mathbf{u}^2}{2f_1} + p_j\left[\frac{k}{f_1}\int_{u_0}^{u} x[\xi(\mathbf{u})]\,\mathrm{d}u + \frac{k}{f_1}\int_{v_0}^{v} y[\xi(\mathbf{u})]\,\mathrm{d}v\right] + \frac{k}{f_1}\mathbf{X}_j\mathbf{u} \tag{6.28}$$

where

$$p_j = 1 + j\frac{f_1}{f_2} \tag{6.29}$$

$$\mathbf{X}_j = \frac{\mathbf{x}_1 f_2 + j\mathbf{x}_2 f_1}{f_2} \tag{6.30}$$

Thus, the multifocus focusator defined by Eqs (6.14)–(6.16) and (6.27) provides focusing into N lines in a single focal plane $z = f_1$ with the power distribution $I_1, \ldots, I_N$. The scale of the focal lines is given by the linear function in Eq. (6.29).

The developed method features an analytically iterative approach to the calculation of multifocus DOEs. The calculation of the functions $\varphi_1(\mathbf{u})$ and $\varphi_2(\mathbf{u})$ in Eqs (6.14) and (6.15) is conducted using analytical formulae (6.2) to (6.8). The calculation of the nonlinear predistortion function $\Phi[\xi]$ corresponding to the multiorder diffraction grating is based on numerical iterative techniques.

6.1.3 Numerical Examples

As an example of the developed method, we consider computing a multifocus focusator specified in Eqs (6.14)–(6.16) and (6.26), and generating three straight lines of equal size located in three different focal planes. The power ratios between the lines are supposed to be constant. For that purpose, we first define the functions $\varphi_1(\mathbf{u})$ and $\varphi_2(\mathbf{u})$ in Eqs (6.14) and (6.15) as the phase functions of the focusators in the straight lines $|x| \leq d$ located in the focal planes $z = f_1$ and $z = f_2$. For the case of a square aperture $G(2a \times 2a)$ and uniform illuminating beam, the functions $\varphi_1(\mathbf{u})$ and $\varphi_2(\mathbf{u})$ can be derived from Eqs (6.6) and (6.7) in the following form

$$\begin{cases} \varphi_1(\mathbf{u}) = -\dfrac{k\mathbf{u}^2}{2f_1} + \dfrac{kd}{2f_1 a}u^2 \\[2ex] \varphi_2(\mathbf{u}) = -\dfrac{k\mathbf{u}^2}{2f_2} + \dfrac{kd}{2f_2 a}u^2 \end{cases} \tag{6.31}$$

Furthermore, we define the nonlinear predistortion function $\Phi[\xi]$ as the phase function of a three-order diffraction grating

$$\Phi[\xi] = \begin{cases} 0, & \xi \in [0, \pi] \\ 2\tan^{-1}(\pi/2), & \xi \in [\pi, 2\pi] \end{cases} \tag{6.32}$$

The Fourier coefficients of the function $\exp(i\Phi[\xi])$ are

$$c_j = \begin{cases} \dfrac{1-(-1)^j}{2\pi i j}\left[1 - \exp\left(2i\tan^{-1}\dfrac{\pi}{2}\right)\right], & j \neq 0, j = \pm 1, \pm 2, \ldots \\ \dfrac{1}{2}\left[1 + \exp\left(2i\tan^{-1}\dfrac{\pi}{2}\right)\right], & j = 0 \end{cases} \tag{6.33}$$

and $|c_{-1}|^2 = |c_0|^2 = |c_1|^2 = 0.2884$. The energy efficiency is then >86%. The substitution of Eqs (6.31) and (6.33) into Eq. (6.20) with the main $j = -1, 0, 1$ terms yields focusing in three straight lines in the planes $z = F_j$, $j = -1, 0, 1$. The centres of lines in these planes are located at points $\mathbf{X}_j$, $j = -1, 0, 1$ defined in Eq. (6.24).

The field intensity formed by the multifocus focusator with the phase function reduced to modulo 2π given by Eqs (6.14)–(6.16), (6.26), (6.31) and (6.32) has been simulated with the following parameters: $\lambda = 1.06\,\mu$m, $2d = 8$ mm, $2a = 9$ mm, $f_1 = 100$ mm, $f_2 = 3000$ mm, $\mathbf{x}_1 = (0, 0)$ mm, and $\mathbf{x}_2 = (145, 0)$ mm. Under these parameters, coordinates of the focal planes are $F_1 = 96.774$ mm, $F_0 = 100$ mm, and $F_{-1} = 103.448$ mm and the centres

Figure 6.4 Portion of zone plate $\Phi[\varphi_d(\mathbf{u})]$

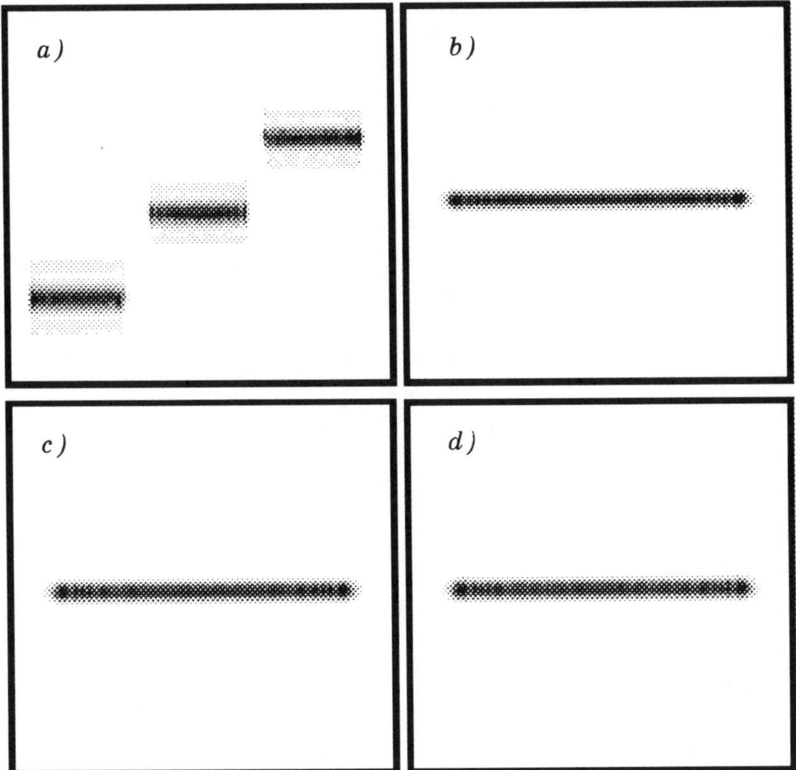

Figure 6.5 Half-tone intensity distributions formed by a multifocus focusator focusing into three lines located in three different focal planes: (a) in the plane *XOZ* containing the focal lines; (b), (c), (d) in the focal planes $z = F_j$, $j = -1, 0, 1$

of lines are shifted from the optical axis in the points $\mathbf{X}_1 = (4.677, 0)$ mm, $\mathbf{X}_0 = (0, 0)$ mm, and $\mathbf{X}_{-1} = (-5, 0)$ mm. Figure 6.4 displays a portion of the zone plate $\Phi[\varphi_d(\mathbf{u})]$. Figure 6.5 illustrates the intensity distributions formed using the multifocus focusator. The plots have been obtained by the numerical computation of the paraxial Kirchhoff integral. Figure 6.5a demonstrates the half-tone intensity distribution in the plane *XOZ* containing the focal lines. The half-tone intensity distributions formed in the focal planes $z = F_j$, $j = -1, 0, 1$ are shown in Figs 6.5b, 6.5c and 6.5d, respectively. The results of computation confirm the robustness of the presented method.

As a second example, let us consider the calculation of a multifocus focusator, Eqs (6.14)–(6.16) and (6.27), with a square aperture $G(2a \times 2a)$ focusing a uniform beam in five scaled straight lines in a single focal plane $z = f_1$. Let 1:2:3:4:5 be the required focal line length ratios (see Fig. 6.6a). The required scale factor is given by Eq. (6.29) with $f_2 = 3f_1$ and $j = -2, 2$. According to Eq. (6.30), the predetermined space x_0 between the focal lines

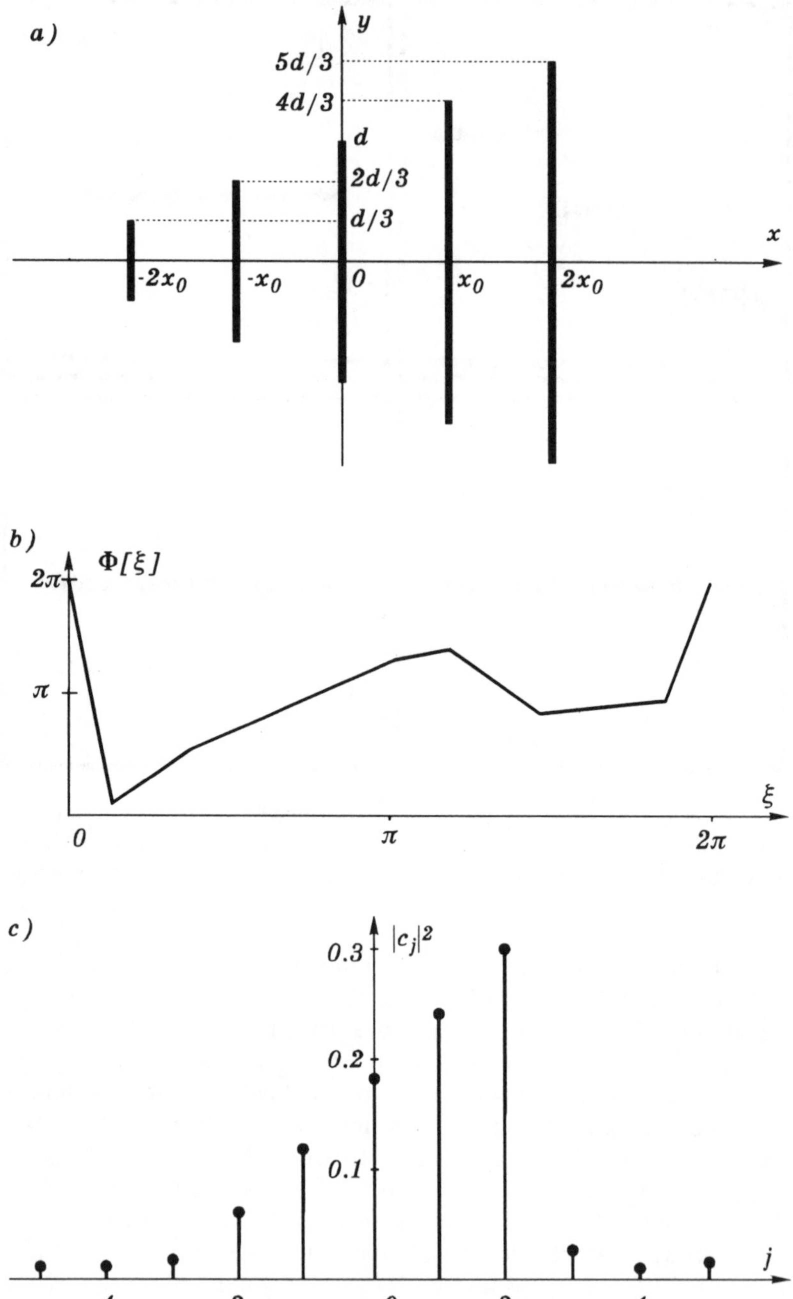

Figure 6.6 (a) Geometry of the segments of focusing; (b) nonlinear predistortion function $\Phi[\xi]$ upon focusing into a set of line segments shown in (a); (c) squared modulus of Fourier coefficients for the function $\exp(i\Phi[\xi])$

takes place under $\mathbf{x}_1 = (0, 0)$ and $\mathbf{x}_2 = (3x_0, 0)$. The functions $\varphi_1(\mathbf{u})$ and $\varphi_2(\mathbf{u})$ in Eqs (6.14) and (6.15) have the form

$$\begin{cases} \varphi_1(\mathbf{u}) = -\dfrac{k\mathbf{u}^2}{2f_1} + \dfrac{kd}{2f_1 a} v^2 \\ \varphi_2(\mathbf{u}) = -\dfrac{k\mathbf{u}^2}{6f_1} + \dfrac{kd}{6f_1 a} v^2 \end{cases} \tag{6.34}$$

Finally, to obtain a phase function of a multifocus focusator, it is necessary to define the nonlinear predistortion function $\Phi[\xi]$ in Eq. (6.14) as the phase function of a five-order diffraction grating. To ensure an equal intensity along all focal lines, the intensities of diffraction orders I_j in the grating design were chosen to be proportional to the lines size

$$I_j/I_0 = 1 + j/3, \quad j = \overline{-2, 2} \tag{6.35}$$

The diffraction grating in Eq. (6.35) was calculated using the AA algorithm described in Chapter 5. The grating energy efficiency was found to be 88.8%, with the r.m.s. error in the formation of the pregiven linear intensity distribution of Eq. (6.35) being 1.4%. The phase function of that grating on a period is shown in Fig. 6.6b. Figure 6.6c represents the squared modules of Fourier coefficients for the grating in Fig. 6.6b. The field intensity formed by the multifocus focusator, given by Eqs (6.14)–(6.16), (6.27) and (6.34) with the function $\Phi[\xi]$ shown in Fig. 6.6b, has been simulated with the following parameters: $\lambda = 1.06\,\mu\text{m}$, $2d = 0.6\,\text{mm}$, $2a = 10\,\text{mm}$, $f_1 = 100\,\text{mm}$, $f_2 = 3f_1 = 300\,\text{mm}$, $\mathbf{x}_1 = (0, 0)\,\text{mm}$, and $\mathbf{x}_2 = (0.75, 0)\,\text{mm}$. The calculated intensity distributions generated by the multifocus element with the above parameters are shown in Fig. 6.7. The result of calculation demonstrates high focusing quality.

6.2 Multifocus Zone Plates and Nonlinear Combined Effects Diffractive Elements

6.2.1 *Multifocus Zone Plates*

If the nonlinear predistortion $\Phi[\xi]$ in Eq. (6.14) is defined as a multiorder binary grating, the phase function

$$\varphi_{zp}(\mathbf{u}) = \Phi[\varphi_d(\mathbf{u})] \tag{6.36}$$

corresponds to a binary zone plate. The zone plate specified in Eqs (6.36) and (6.15) is a particular case of the multifocus focusator of Eqs (6.14) and (6.15). Since binary elements are of particular interest because of their easy manufacturability, we pay particular attention to multifocus properties of the zone plate $\varphi_{zp}(\mathbf{u})$. Assume that the zone plate in Eqs (6.36) and (6.15)

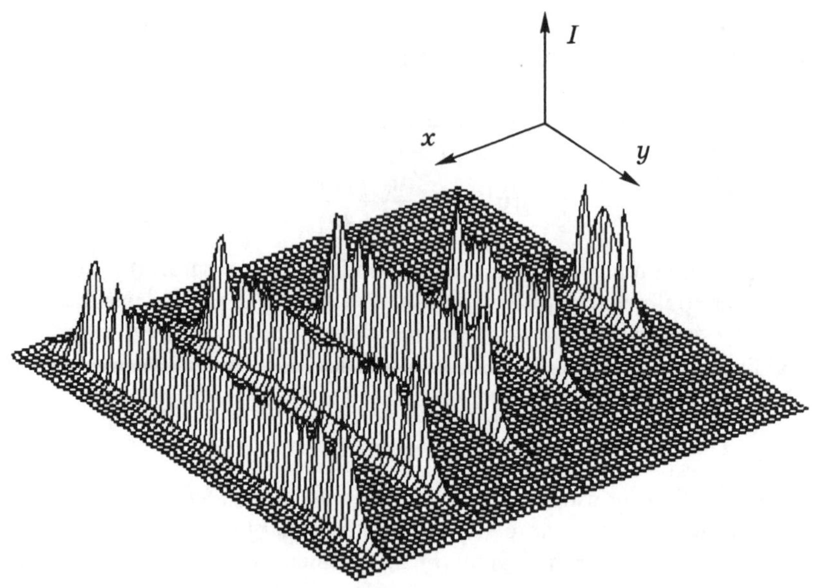

Figure 6.7 Intensity distribution produced by a multifocus focusator in plane $z = f_1$ upon focusing into a set of line segments shown in Fig. 6.6a

is illuminated by a convergent spherical beam with the complex amplitude given by

$$W_0(\mathbf{u}) = \sqrt{[I_0(\mathbf{u})]}\exp\left(-\frac{ik\mathbf{u}^2}{2f_1}\right) \tag{6.37}$$

Using Eqs (6.15) and (6.18), we represent the field immediately beyond the zone plate $\varphi_{zp}(\mathbf{u})$ in the form

$$\begin{aligned} W(\mathbf{u}) &= \sqrt{[I_0(\mathbf{u})]}\exp\left(-\frac{ik\mathbf{u}^2}{2f_1}\right)\sum_{j=-\infty}^{\infty} c_j \exp[ij\varphi_d(\mathbf{u})] \\ &= \sqrt{[I_0(\mathbf{u})]}\sum_{j=-\infty}^{\infty} c_j \exp[i\varphi_j(\mathbf{u})] \end{aligned} \tag{6.38}$$

where c_j are the Fourier coefficients given in Eq. (6.19). The functions $\varphi_j(\mathbf{u})$ in Eq. (6.38), defining the structure of the images formed in the diffraction orders of the zone plate $\varphi_{zp}(\mathbf{u})$, have the form of Eq. (6.21) and correspond to the phase functions of a focusator focusing into a scaled line L. The parameters F_j, p_j, and $\mathbf{X}_j$ defining the scale and location of the focal line formed in the jth diffraction order are given by

$$F_j = \frac{f_1 f_2}{f_2 + jf_1(1 + f_2/f_3)} \tag{6.39}$$

$$p_j = \frac{jf_2}{f_2 + j(1 + f_2/f_3)} \tag{6.40}$$

$$\mathbf{X}_j = p_j \mathbf{x}_2 \tag{6.41}$$

In the case of $f_3 = -f_2$, the focal lines locate in a single focal plane $z = f_1$. The scale of lines is given by the linear function

$$p_j = jf_1/f_2 \tag{6.42}$$

Thus, the zone plate of Eqs (6.36) and (6.15) with the function $\Phi[\xi]$ defined as the phase function of an N-order diffraction grating with intensities $I_1, \ldots, I_N$ in the orders $j_1, \ldots, j_N$ provides the focusing of the convergent spherical beam of Eq. (6.37) into N scaled lines with power ratios $I_1, \ldots, I_N$ and located in the planes $z = F_{j_i}$, $i = 1, N$.

For the function $\Phi[\xi]$ defined as the phase function of the N-order binary grating, modules of the Fourier coefficients c_j obey the equation (see Eqs (5.14) and (5.15))

$$|c_j| = |c_{-j}|, \quad j = -\infty, \infty \tag{6.43}$$

Due to this symmetry, the design of the multifocus binary zone plate can be performed only for the symmetric set of indices j_i, the desired power ratios $I_1, \ldots, I_N$ between the focal lines also need to be symmetric.

6.2.2 *Binary Zone Plates for Focusing into Symmetric Focal Contours*

For $f_3 = -f_2$, the zone plate specified by Eqs (6.36) and (6.15) focuses the convergent spherical beam of Eq. (6.37) into a set of lines L_{p_j} in the plane $z = f_1$. According to Eqs (6.8), (6.9), and (6.39)–(6.41), the focal lines L_{p_1} and $L_{p_{-1}}$ generated in the 1st and -1st diffraction orders of the zone plate $\varphi_{zp}(\mathbf{u})$ are centrally symmetric.

Let us consider the calculational procedure for the binary zone plate dedicated to focusing into a centrally symmetric contour C in the plane $z = f_1$. The contour C can be represented as a combination of two centrally symmetric lines. Let us assume that the $\varphi_d(\mathbf{u})$ function in Eq. (6.36) is the phase function of a focusator focusing into a line L equal to one-half the focal contour C. In that case, the focal lines $L_{p_1} \equiv L$ and $L_{p_{-1}}$ formed in the 1st and -1st diffraction orders of the zone plate $\varphi_{zp}(\mathbf{u})$ produce the desired focal contour C. For the purpose of calculating the binary zone plate $\varphi_{zp}(\mathbf{u})$, we define the function $\Phi[\xi]$ as the phase function of a binary two-order grating

$$\Phi[\xi] = \begin{cases} 0, & \xi \in [0, \pi] \\ \pi, & \xi \in [\pi, 2\pi] \end{cases} \tag{6.44}$$

The Fourier coefficients c_j in the expansion of the function $\exp(i\Phi[\xi])$ are given by

$$c_j = \begin{cases} \dfrac{1-(-1)^j}{\pi i j}, & j = \pm 1, \pm 2, \ldots \\ 0, & j = 0 \end{cases} \tag{6.45}$$

assuming that $|c_1|^2 = |c_{-1}|^2 = 0.405$. Thus, having regard to the basic 1st and -1st terms of the expansion in Eq. (6.38), the binary zone plate of Eqs (6.36) and (6.44) focuses 81% of the illuminating beam energy into the lines L_{p_1} and $L_{p_{-1}}$, producing the desired focal contour C.

When the line L has a complex, non-smooth, or disconnected configuration, the calculation of the focusator phase function is found to be an arduous or even impossible task. In such a situation, the $\varphi_d(\mathbf{u})$ function can be deduced in a Fresnel approximation using the GS algorithm. When the $\varphi_d(\mathbf{u})$ function in Eq. (6.15) is calculated using Eqs (6.2) and (6.3) obtained within the ray-tracing approach, we find the line $L_{p_{-1}}$ generated in the -1st order of the zone plate $\varphi_{zp}(\mathbf{u})$. In the Fresnel approximation, the intensities of fields formed by the zone plate $\varphi_{zp}(\mathbf{u})$ in the 1st and -1st orders will be given by

$$I_1(x) = \left| \frac{c_1}{\lambda f} \int_G \sqrt{[I_0(\mathbf{u})]} \exp\left(i\varphi_d(\mathbf{u}) - \frac{ik}{f}\mathbf{xu} \right) \mathrm{d}^2\mathbf{u} \right|^2 \tag{6.46}$$

$$I_{-1}(x) = \left| \frac{c_{-1}}{\lambda f} \left[\int_G \sqrt{[I_0(\mathbf{u})]} \exp\left(i\varphi_d(\mathbf{u}) - \frac{ik}{f}(-\mathbf{x})\mathbf{u} \right) \mathrm{d}^2\mathbf{u} \right]^* \right|^2 \tag{6.47}$$

where * symbolizes complex conjugation.

According to Eqs (6.46) and (6.47), we have

$$I_{-1}(\mathbf{x}) = I_1(-\mathbf{x}) \tag{6.48}$$

Hence, if $I_1(\mathbf{x})$ corresponds to the field intensity upon focusing into the line $L_{p_1} \equiv L$, in order -1 we shall have the line L_{p-1}. The result arrived at shows that in calculating the zone plates of Eq. (6.36) intended for the formation of focal lines only in the 1st and -1st diffraction orders, the $\varphi_d(\mathbf{u})$ function can be derived via iteration methods.

6.2.3 *Nonlinear Combined Effects Diffractive Elements*

Linear combined effects of a lens and a grating can be generated by a single diffractive element whose phase function is an ordinary sum of the lens phase and the grating phase. The superposition of the complex transmission functions of any two binary elements represented as

$$T(\mathbf{u}) = \frac{1}{\sqrt{2}} \{\exp[i\varphi_{b1}(\mathbf{u})] + i\exp[i\varphi_{b2}(\mathbf{u})]\} \tag{6.49}$$

corresponds to the purely phase element with the phase function given by [121]

$$\varphi[\varphi_{b1}(\mathbf{u}), \varphi_{b2}(\mathbf{u})] = \begin{cases} \pi/4 & \varphi_{b1}(\mathbf{u}) = 0, \quad \varphi_{b2}(\mathbf{u}) = 0 \\ 3\pi/4 & \varphi_{b1}(\mathbf{u}) = \pi, \quad \varphi_{b2}(\mathbf{u}) = 0 \\ 5\pi/4 & \varphi_{b1}(\mathbf{u}) = \pi, \quad \varphi_{b2}(\mathbf{u}) = \pi \\ 7\pi/4 & \varphi_{b1}(\mathbf{u}) = 0, \quad \varphi_{b2}(\mathbf{u}) = \pi \end{cases} \tag{6.50}$$

We use this property to design a kind of 'nonlinear combined effects element' which focuses in two sets of lines simultaneously. For the purpose of our design, we introduce the function $\varphi(\varphi_{zp1}(\mathbf{u}), \varphi_{zp2}(\mathbf{u}))$ of Eq. (6.50), corresponding to the superposition, Eq. (6.49), of the binary multifocus zone plates $\varphi_{zp1}(\mathbf{u})$ and $\varphi_{zp2}(\mathbf{u})$ focusing the converging spherical beam of Eq. (6.37) in two sets S_1 and S_2 of lines, respectively.

We propose that the phase function of the combined effects element should be defined in the form

$$\varphi_{com}(\mathbf{u}) = \varphi_l(\mathbf{u}) + \varphi[\varphi_{zp1}(\mathbf{u}), \varphi_{zp2}(\mathbf{u})] - \varphi_0(\mathbf{u}), \quad \mathbf{u} \in G \tag{6.51}$$

where $\varphi_0(\mathbf{u})$ is the phase of the illuminating beam and the function

$$\varphi_l(\mathbf{u}) = -\frac{k\mathbf{u}^2}{2f_1}$$

is the paraxial phase of the lens.

According to Eqs (6.51) and (6.49), the field immediately beyond the plane of the combined effects element illuminated by the beam with the complex amplitude given by

$$W_0(\mathbf{u}) = \sqrt{[I_0(\mathbf{u})]}\,\mathrm{e}^{i\varphi_0(\mathbf{u})}$$

takes the form

$$W(\mathbf{u}) = \sqrt{\left(\frac{I_0(\mathbf{u})}{2}\right)} \exp\left(-\frac{ik\mathbf{u}^2}{2f_1} + i\varphi_{zp1}(\mathbf{u})\right) + i\sqrt{\left(\frac{I_0(\mathbf{u})}{2}\right)} \exp\left(-\frac{ik\mathbf{u}^2}{2f_1} + i\varphi_{zp2}(\mathbf{u})\right) \tag{6.52}$$

The terms of Eq. (6.52) correspond (up to a constant) to the fields formed by the binary zone plates $\varphi_{zp1}(\mathbf{u})$ and $\varphi_{zp2}(\mathbf{u})$ under the converging spherical beam of Eq. (6.37). According to the linearity of the space propagation operator, the combined effects element, Eq. (6.51), can focus simultaneously into a set of lines S_1 and into a set of lines S_2.

6.2.4 Numerical Examples

Let us consider the calculation of the binary plate specified by Eqs (6.36) and (6.15) and aimed at focusing the converging spherical beam of Eq. (6.37) into a set of six line segments with length ratios of 3:2:1:1:2:3 in the plane

$z=f_1$ (see Fig. 6.8a). According to Eqs (6.39) to (6.41), focusing into the desired set of line segments takes place under the following parameters: $f_1=f_2$, $f_3=-f_2$ and $\mathbf{x}_2=(x_2,0)$. Note that the $\varphi_2(\mathbf{u})$ function in Eq. (6.15) is the phase function of a focusator focusing into the line segment $|x|\leq d$ in the plane $z=f_1$. Given the square aperture $G(2a\times 2a)$ and a uniform intensity of illuminating beam, the function $\varphi_2(\mathbf{u})$ can be deduced from Eqs (6.6) and (6.7) in the form

$$\varphi_2(\mathbf{u}) = -\frac{k\mathbf{u}^2}{2f_1}+\frac{kd}{2f_1 a}v^2 \tag{6.53}$$

To form a uniform intensity along focal lines, the $\Phi[\xi]$ function should be defined as the phase function of a binary six-order grating characterized by order intensities proportional to the length of segments of focusing

$$I_j/I_0=|j|,\quad j=\overline{-3,3}\quad j\neq 0 \tag{6.54}$$

The diffraction grating of Eq. (6.54) was designed using the gradient method discussed in Chapter 5. The grating energy efficiency was found to be 74.5%, with the r.m.s. error in the formation of the required linear intensity distribution, Eq. (6.54), being 1.6%.

The calculated binary profile of a grating period is shown in Fig. 6.8b, with values of the squared modules of the Fourier coefficients, Eq. (6.19), depicted in Fig. 6.8c. Figure 6.9 illustrates the calculated intensity distribution synthesized by the zone plate of Eqs (6.36), (6.15), and (6.53), with the $\Phi[\xi]$ shown in Fig. 6.8b for the following parameters: $\lambda=1.06\,\mu$m, $2d=0.5$ mm, $2a=10$ mm, $f_1=f_2=-f_3=100$ mm, and $\mathbf{x}_2=(0.25,0)$ mm. Figure 6.9 shows high quality of focusing; however, there exists a sharp intensity peak in the centre, which is due to the error in the calculation of the function $\Phi[\xi]$. The Fourier coefficient c_0 in the expansion of the function $\exp(i\Phi[\xi])$ is non-zero; $|c_0|^2\approx 0.01$.

According to Eq. (6.42), an image generated in the zero order corresponds to a point. Since in the desired orders $j=\overline{-3,3}, j\neq 0$, the energy is distributed uniformly over the focal segment length, the central peak intensity is found to be large. At the same time, the portion of illuminating beam energy focused in the central peak amounts only to 1%.

As a second example, we consider the calculation of a binary zone plate expressed by Eqs (6.36) and (6.44) and aimed to focus into the letter 'x' in the plane $z=f_1$. The letter 'x' is assumed to be composed of two centrally symmetric semirings of radius R_0. In our case, the function $\varphi_d(\mathbf{u})$ of Eq. (6.36) is the phase function of a focusator of the convergent spherical beam focusing into a semiring centred at point $(-x_0,0)$ in the plane $z=f_1$. The parameter x_0 determines the spacing between the semirings. With a circular illuminating beam of uniform intensity, the $\varphi_d(\mathbf{u})$ function takes the form [136]

$$\varphi_d(\mathbf{u}) = \mathrm{mod}_{2\pi}\left[\frac{k}{f_1}R_0\,\mathrm{sign}(u)|\mathbf{u}|-\frac{k}{f_1}x_0 u\right] \tag{6.55}$$

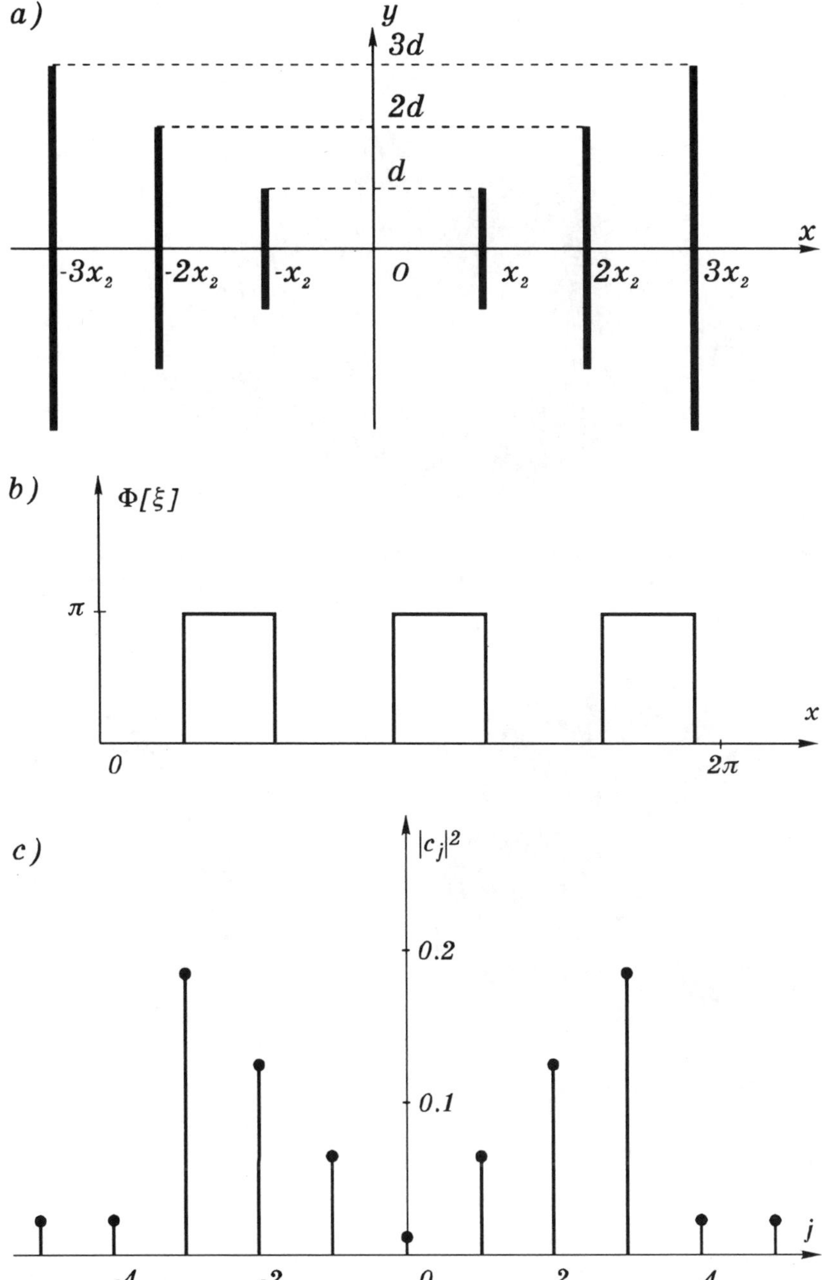

Figure 6.8 (a) Geometry of the segments of focusing; (b) binary function $\Phi[\xi]$ of nonlinear predistortion upon focusing into the set of line segments shown in (a); (c) squared modules of Fourier coefficients for the function $\exp(i\Phi[\xi])$

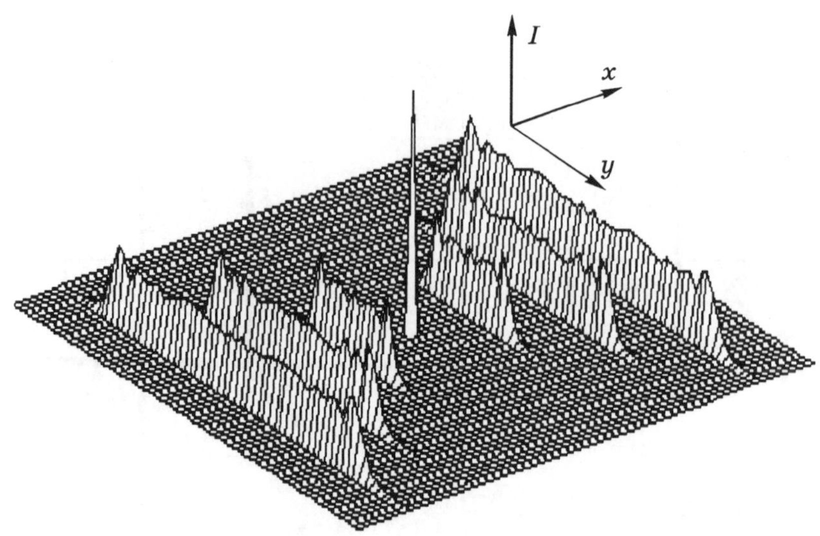

Figure 6.9 Intensity distribution produced by a multifocus binary zone plate in plane $z = f_1$ upon focusing into the set of line segments shown in Fig. 6.8a

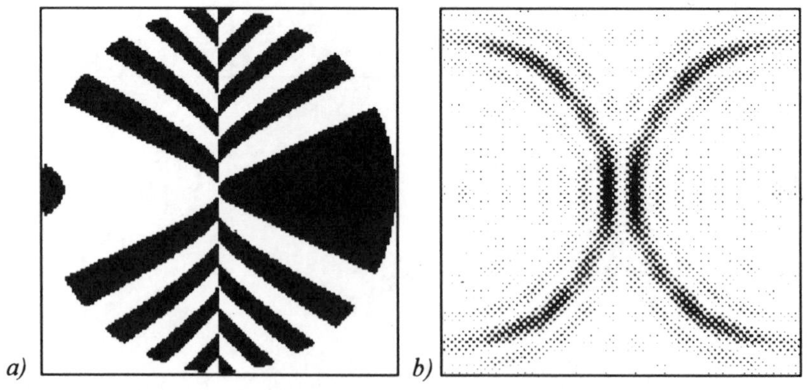

Figure 6.10 (a) Photomask of a binary zone plate for focusing into the letter 'x'; (b) half-tone intensity distribution produced in plane $z = f_1$ by the binary zone plate shown in (a)

where

$$\text{sign}(u) = \begin{cases} 1, & u \geq 0 \\ -1, & u < 0 \end{cases}$$

Figure 6.10a illustrates the zone plate of Eqs (6.36), (6.44), and (6.55) calculated for the following parameters: $\lambda = 10.6\,\mu\text{m}$, $R_0 = 5\,\text{mm}$,

$f_1 = 500\,\text{mm}$, $x_0 = 5.5\,\text{mm}$, and the radius of the illuminating beam is $R = 4\,\text{mm}$. To avoid interference between the fields generated in the 1st and -1st orders, we put $x_0 > R_0$. The calculated half-tone intensity distribution given in Fig. 6.10b corroborates the serviceability of the zone plate.

In conclusion, let us consider the calculation of the combined effects element of Eq. (6.51) intended for the focusing of a circular beam with uniform intensity into a square contour of size $2d \times 2d$. The square contour can be represented as two sets of line segments S_1 and S_2. Set S_1 consists of two centrally symmetric line segments, viz. $x = d$, $|y| \leq d$ and $x = -d$, $|y| \leq d$, parallel to the axis Oy. Set S_2 includes the centrally symmetric line segments $y = d$, $|x| \leq d$ and $y = -d$, $|x| \leq d$, parallel to the axis Ox. In this case, $\varphi_{zp1}(\mathbf{u})$ and $\varphi_{zp2}(\mathbf{u})$ in Eq. (6.51) are binary zone plates for focusing in the plane $z = f_1$ into the sets S_1 and S_2, comprising centrally symmetric line segments. By virtue of symmetry of the problem of focusing, we have

$$\varphi_{zp1}(u, v) = \varphi_{zp2}(v, -u) \tag{6.56}$$

The phase function of a zone plate aimed at focusing into the set of line segments S_1 takes the form of Eqs (6.36) and (6.44).

The function $\varphi_d(\mathbf{u})$ in Eq. (6.36) takes the form of Eq. (6.15) given that $f_1 = f_2 = -f_3$, $\mathbf{x}_2 = (0, d)$, and the function $\varphi_2(\mathbf{u})$ is the phase function of a focusator focusing into a segment $|x| \leq d$, in the plane $z = f_1$. With a circular illuminating beam (radius R) of uniform intensity, the $\varphi_2(\mathbf{u})$ function can be derived from Eqs (6.6) and (6.7) in the form

$$\varphi_2(\mathbf{u}) = -\frac{k\mathbf{u}^2}{2f_1} + \frac{2kd}{\pi f_1} \times \left\{ u \arcsin\left(\frac{u}{R}\right) + R\left[1 - \left(\frac{u}{R}\right)^2\right]^{1/2} - \frac{R}{3}\left[1 - \left(\frac{u}{R}\right)^2\right]^{3/2} \right\} \tag{6.57}$$

Since for the binary grating of Eq. (6.44) we have $|c_1|^2 = |c_{-1}|^2 = 0.405$, according to the expansion in Eq. (6.38) and Eqs (6.39)–(6.41), the zone plate specified by Eqs (6.36), (6.15), (6.44), and (6.57) can focus 81% of the illuminating beam energy into the desired set of line segments S_1. The field intensity formed by the combined effects element with the phase function reduced to modulo 2π given by Eqs (6.51), (6.56), (6.57), (6.15), and (6.44) has been simulated with the following parameters: $\lambda = 10.6\,\mu\text{m}$, $2d = 8\,\text{mm}$, $R = 5\,\text{mm}$, and $f_1 = f_2 = -f_3 = 350\,\text{mm}$.

Figure 6.11a displays an amplitude mask of the zone plate $\varphi[\varphi_{zp1}(\mathbf{u}), \varphi_{zp2}(\mathbf{u})]$ in Eq. (6.51). Figure 6.11b illustrates the half-tone intensity distributions formed by the combined effects element, thus confirming high robustness of the element developed.

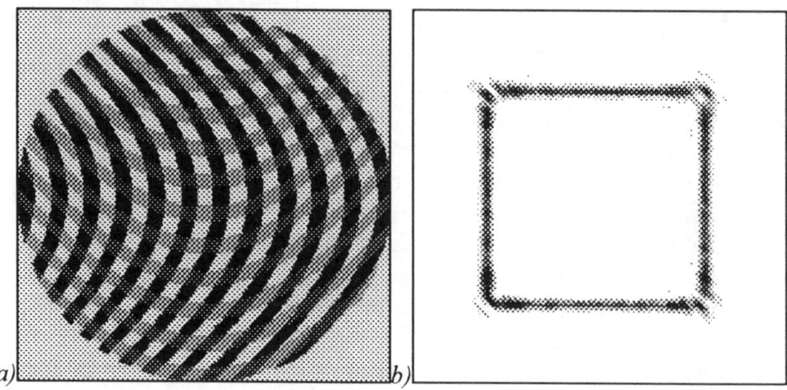

Figure 6.11 (a) Photomask of the zone plate $\varphi[\varphi_{zp1}(\mathbf{u}), \varphi_{zp2}(\mathbf{u})]$ upon focusing into a square contour; (b) half-tone intensity distribution produced in plane $z = f_1$ by the nonlinear combined effects diffractive element

6.3 Diffractive Multifocus Lens

As a special case of the previous method, we consider the computation of a multifocus diffractive lens providing focusing into N points on the optical axis with power distribution $I_1, \ldots, I_N$. The point is a particular case of a focal line. Then we replace in Eq. (6.14) the phase functions of the focusator focusing into a line by the phase functions of the lens. Hence, the phase function of the multifocus lens is represented in the form [47,140,141]

$$\varphi_{ml}(\mathbf{u}) = \varphi_1(u) + \Phi[\varphi_2(\mathbf{u})] - \varphi_0(\mathbf{u}), \quad \mathbf{u} \in G \tag{6.58}$$

where

$$\begin{aligned} \varphi_1(\mathbf{u}) &= -\frac{k\mathbf{u}^2}{2f_1} \\ \varphi_2(\mathbf{u}) &= \mathrm{mod}_{2\pi}\left(-\frac{k\mathbf{u}^2}{2f_2}\right) \end{aligned} \tag{6.59}$$

are the paraxial phases of the lenses with focal lengths f_1 and f_2. The function $\Phi[\varphi_2(\mathbf{u})]$ in Eq. (6.58) describes a nonlinear predistortion of the lens phase $\varphi_2(\mathbf{u})$ (see Fig. 6.3). Thus, the multifocus lens is expressed as a mathematical superposition of the conventional lens $\varphi_1(\mathbf{u})$ and the zone plate $\Phi[\varphi_2(\mathbf{u})]$.

According to the general representation in Eq. (6.20), the field immediately beyond the plane of the multifocus lens illuminated by the beam with complex amplitude

$$W_0(\mathbf{u}) = \sqrt{[I_0(\mathbf{u})]}\, e^{i\varphi_0(\mathbf{u})}$$

is given by

$$W(\mathbf{u}) = \sqrt{[I_0(\mathbf{u})]}\exp\left(-\frac{ik\mathbf{u}^2}{2f_1}\right)\sum_{j=-\infty}^{\infty} c_j \exp\left[ij\left(-\frac{k\mathbf{u}^2}{2f_2}\right)\right] \quad (6.60)$$

The expression in Eq. (6.60) corresponds to the superposition of a set of paraxial spherical beams with focal lengths

$$F_j = \frac{f_1 f_2}{f_2 + jf_1} \quad (6.61)$$

and power distribution $|c_j|^2$, $j = -\infty, \infty$ between them. The Fourier coefficients c_j of the function $\exp(i\Phi[\xi])$ are defined in Eq. (6.19). Let $j_1, \ldots, j_N$ be the indices corresponding to the required focal lengths. Following the previous considerations, it will suffice to determine the nonlinear predistortion $\Phi[\xi]$ as the phase function of an N-order diffraction grating with intensities $I_1, \ldots, I_N$ of the orders $j_1, \ldots, j_N$.

The design of the diffraction grating $\Phi[\xi]$ calls for appropriate numerical techniques. To compute the nonlinear predistortion function $\Phi[\xi]$ in Eq. (6.58), we can employ the analytical method for computing 1D focusators [37,126]. Let $\tilde{\varphi}(\xi)$ $0 < \xi < d$ be the phase function of a 1D focusator that focuses the converging cylindrical beam into a segment with an intensity distribution of $I(x)$, for $n_1\Delta < x < n_2\Delta$, where $\Delta = \lambda f/d$ is the size of the diffraction spot. The function $\tilde{\varphi}(\xi)$ has the form [37,126]

$$\tilde{\varphi}(\xi) = \frac{k}{f}\int_0^{\xi} \chi(\eta)\,\mathrm{d}\eta \quad (6.62)$$

where the function $\chi(\eta)$ can be found from the solution of the differential equation

$$\frac{\mathrm{d}\chi(\xi)}{\mathrm{d}\xi} = \frac{1}{I[\chi(\xi)]} \quad (6.63)$$

with the boundary conditions $\chi(\eta) = n_1\Delta, \chi(\eta) = n_2\Delta$.

It is shown in section 5.1 that the function $\tilde{\varphi}(\xi)$ can be treated as the phase function of a diffraction grating with a period of d and the intensity in diffraction orders given by

$$I_j = \begin{cases} I(j\Delta), & j = \overline{n_1, n_2} \\ 0, & \text{else} \end{cases} \quad (6.64)$$

Thus, we can find the function of the nonlinear predistortion $\Phi[\xi] = \tilde{\varphi}(\xi)$ from Eqs (6.62) and (6.63). As a result of our investigations, we have found that the ray-tracing approach used to compute the function $\Phi[\xi] = \tilde{\varphi}(\xi)$ gives reliable results only for a large number of orders (30–40 and above).

6.3.1 Combined Effects Multifocus Lenses

In this section we consider 'combined effects multifocus lenses' able to focus simultaneously in two sets of points; the first set consists of the points located in one plane and the second of the points located on the optical axis. These lenses are a special case of combined effects diffractive elements considered in section 6.2.3.

If the nonlinear predistortion $\Phi[\xi]$ in Eq. (6.58) is defined as a multiorder binary grating, the function $\Phi[\varphi_2(\mathbf{u})]$ corresponds to a binary zone plate. The superposition in Eq. (6.49) of complex transmission functions of the binary zone plate $\Phi[\varphi_2(\mathbf{u})]$ and the multiorder binary grating $\varphi_{gr}(\mathbf{u})$ corresponds to the phase-only element with the phase function $\varphi[\Phi[\varphi_2(\mathbf{u})], \varphi_{gr}(u)]$ given in Eq. (6.50).

We propose that the phase function of the combined effects multifocus lens should be defined in the form

$$\varphi_{cml}(\mathbf{u}) = \varphi_1(\mathbf{u}) + \varphi\{\Phi[\varphi_2(\mathbf{u})], \varphi_{gr}(u)\} - \varphi_0(\mathbf{u}), \quad \mathbf{u} \in G \tag{6.65}$$

where $\varphi_0(\mathbf{u})$ is the phase of the illuminating beam and $\varphi_1(\mathbf{u})$ is the phase function of the lens as used in Eq. (6.58).

According to Eqs (6.49) and (6.50), we can represent the complex transmission function of the lens in Eq. (6.65) in the form

$$\exp[i\varphi_{cml}(\mathbf{u})] = \frac{1}{\sqrt{2}}\exp[i\varphi_{ml}(\mathbf{u})] + \frac{i}{\sqrt{2}}\exp[i\varphi_1(\mathbf{u}) + i\varphi_{gr}(u) - i\varphi_0(\mathbf{u})] \tag{6.66}$$

where the first term corresponds to the multifocus lens of Eq. (6.58) with the nonlinear predistortion $\Phi[\xi]$ defined as a binary grating. The second term contains the phase functions of the lens and of the binary grating, ensuring focusing in a set of points in the plane $z = f_1$. The phase function $\varphi_0(\mathbf{u})$ in the second term compensates for the phase of the illuminating beam. According to the linearity of the space propagation operator, the combined effects lens in Eq. (6.65) can simultaneously focus on a set of points along the optical axis and in one fixed plane.

6.3.2 Computation and Investigation of Multifocus Lenses

As an important example of the method developed, we consider a procedure of synthesizing a bifocal lens with focal lengths of f_a and f_b and with equal powers in each focus. Aperture G is circular with radius R. In this case, in the general expression of Eq. (6.61) for the focal lengths of multifocus lenses we put $F_1 = f_a, F_{-1} = f_b$ and obtain

$$f_1 = \frac{2f_a f_b}{f_a + f_b}, \quad f_2 = \frac{2f_a f_b}{f_a - f_b} \tag{6.67}$$

Furthermore, we define the nonlinear predistortion function $\Phi[\xi]$ as the phase function of a two-order diffraction grating given in Eqs (6.44) and (6.45). The substitution of Eqs (6.67) and (6.45) into Eq. (6.60) with $j = \pm 1$ main non-zero terms yields the following expression for the output field upon the plane illuminating beam

$$W(\mathbf{u}) = c_1 \exp\left(- \frac{ik\mathbf{u}^2}{2f_a} \right) + c_{-1} \exp\left(- \frac{ik\mathbf{u}^2}{2f_b} \right) \tag{6.68}$$

Equation (6.68) describes the required process of focusing into two points (in a paraxial approximation). According to Eq. (6.45), we have $|c_1|^2 = |c_{-1}|^2 = 0.405$. Thus, 81% of the illuminating beam power concentrates in the required foci. We have conducted the computer-aided simulation for the field intensity formed by the diffractive bifocal lens with the phase function of Eqs (6.58), (6.67) and (6.44) and microrelief height taken to the modulus 2π. A segmentized lens with the phase function given by

$$\varphi_s(\mathbf{u}) = \begin{cases} -\dfrac{k\mathbf{u}^2}{2f_a}, & |\mathbf{u}| \leq \dfrac{R}{\sqrt{2}} \\ -\dfrac{k\mathbf{u}^2}{2f_b}, & \dfrac{R}{\sqrt{2}} < |\mathbf{u}| \leq R \end{cases} \tag{6.69}$$

was also under consideration for comparison.

Figure 6.12a illustrates plots of the normalized intensity distribution I along the optical axis z for the diffractive bifocal lens (solid curve) and segmentized lens of Eq. (6.69) (broken curve). The plots have been obtained by the numerical computation of the paraxial Fresnel–Kirchhoff integral using the following parameters: $\lambda = 0.555\ \mu\text{m}$, $f_a = 30$ mm, $f_b = 34$ mm, and $R = 1.5$ mm. Figure 6.12a shows that the depth of focus for the diffractive bifocal lens is less than that for the segmentized lens because the entire aperture operates in each focus of the diffractive lens. Figures 6.12b and 6.12c depict the computer simulated point-spread function on two focal planes. Let us introduce the energy efficiency E as the power inside the spot-size with radius $\Delta = 0.61\lambda z/R$ normalized by the incident beam power. On the first focal plane ($z = 30$ mm) the value of E equals 32.8% and 32.6% for the diffractive and segmentized lens, respectively. On the second focal plane ($z = 34$ mm) E is equal to 34.2% and 14.5%, respectively. Only the central segment of the segmentized lens contributes to the first focal plane. Therefore, the spot-size of the segmentized lens is 1.5 times greater than that of the bifocal diffractive lens (see Fig. 6.12b).

Figure 6.12c depicts a similar effect in the second focal plane where a double-maximum point-spread function occurs for the segmentized lens due to its ring-like pupil. So, we can conclude that the proposed diffractive bifocal lens operates similarly to the conventional lens in each of two foci. Furthermore, its energy efficiency in the second focal plane is twice that of the segmentized lens.

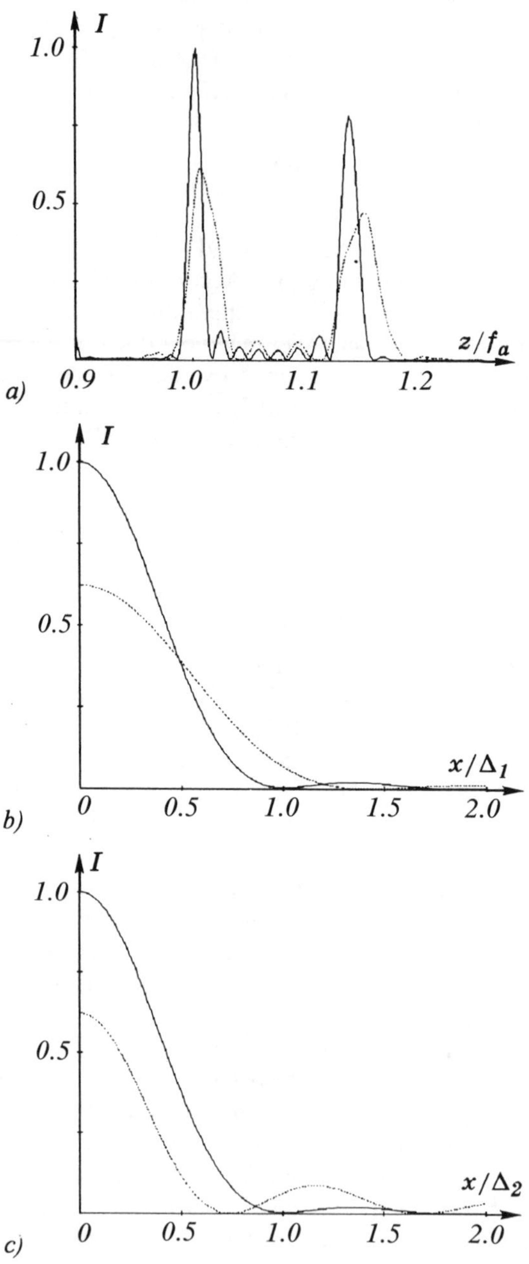

Figure 6.12 Intensity distribution for a diffractive (solid line) and segmentized (broken line) bifocal lens: (a) along the optical axis; (b) and (c) in the focal planes $z = f_a$ and $z = f_b$, where $\Delta_1 = 0.61\lambda f_a/R$ and $\Delta_2 = 0.61\lambda f_b/R$

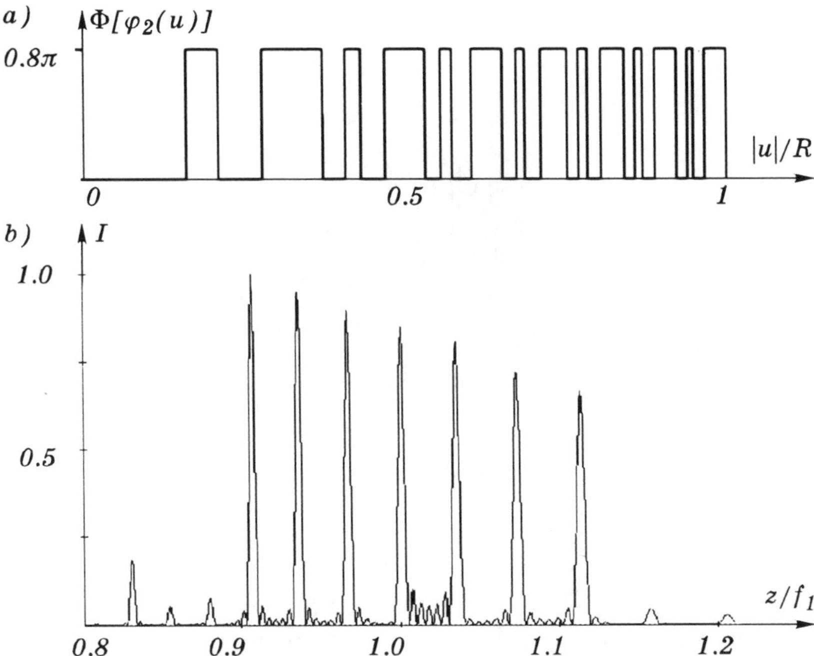

Figure 6.13 (a) Function $\Phi[\varphi_2(\mathbf{u})]$ for a seven-focus lens; (b) axial intensity distribution for the seven-focus lens

The calculation of the lens of Eq. (6.58) with more than two foci presents no problem. The seven-focus lens was calculated as a superposition of the binary zone plate $\Phi[\varphi_2(\mathbf{u})]$ (see Fig. 6.13a) and conventional lens phase according to Eq. (6.58). The zone plate in Fig. 6.13a was computed with the $\Phi[\xi]$ function defined as a seven-order binary diffraction grating. Figure 6.13b illustrates the plot of the normalized intensity distribution I along the optical axis z for the seven-focus lens calculated using the following parameters: $\lambda = 1.06\,\mu$m, $f_1 = 34$ mm, $f_2 = 1000$ mm, and $R = 4$ mm. No less than 80% of illuminating beam power is directed into desired foci. A minor decrease in focal peaks in Fig. 6.13b is due to the expansion of diffraction blurring with increasing focal distance.

The analytical computation of the multifocus lens is possible if we consider the phase function of Eqs (6.62) and (6.63) of a 1D focusator focusing into a segment as the nonlinear predistortion $\Phi[\xi]$. For example, the phase function of the lens with the number of foci $N = n_2 - n_1$ and with equal powers in each focus has the form

$$\varphi_{mf}(\mathbf{u}) = \varphi_1(\mathbf{u}) + \varphi_2(\mathbf{u})\left[n_1 + \frac{n_2 - n_1}{4\pi}\varphi_2(\mathbf{u}) \right] - \varphi_0(\mathbf{u}) \qquad (6.70)$$

The lens in Eq. (6.70) has foci specified by Eq. (6.61), for $j = \overline{n_1, n_2}$.

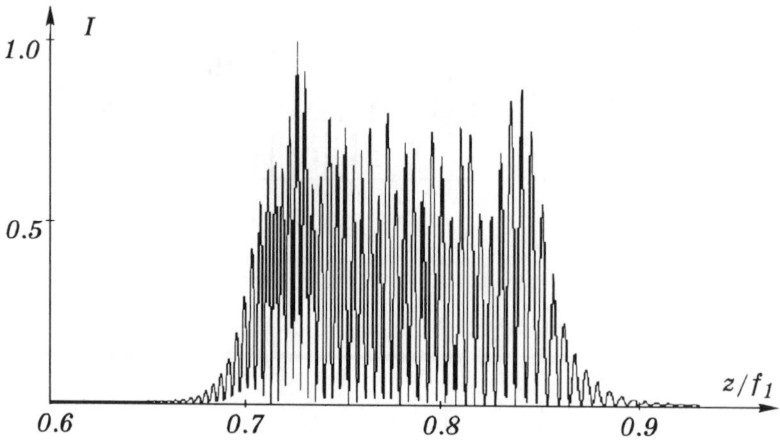

Figure 6.14 Axial intensity distribution for an analytically derived 40-focus lens of Eq. (6.70)

To estimate the efficiency of the analytically computed multifocus lens of Eq. (6.70), we have computed the intensity distribution along the optical axis of this lens with the following parameters: the number of foci $N = 40$ ($n_1 = 20$, $n_2 = 60$), $\lambda = 1.06\,\mu\text{m}$, $f_1 = 30\,\text{mm}$, $f_2 = 4000\,\text{mm}$, and $R = 4\,\text{mm}$. The result of this computation is presented in Fig. 6.14. The nonuniformity of the focal peaks is a consequence of both the ray-tracing approach used in computing the function $\Phi[\xi]$ and the expansion of the diffraction blurring with increasing focal distance. More than 80% of the illuminating beam power is concentrated in the required foci.

According to Eq. (6.60), the multifocus lens of Eq. (6.58) operates similarly to the conventional lens in each focus. If

$$f_2 = \frac{R^2}{2\lambda} \tag{6.71}$$

the lens $\varphi_2(\mathbf{u})$ in Eq. (6.58) has a single central zone. For the multifocus lens of Eqs (6.58) and (6.71), the spacing between adjacent foci

$$l_j = F_{j-1} - F_j \approx 2\lambda \left(\frac{F_j}{R} \right)^2 \tag{6.72}$$

is half as large as the Fresnel length $l_f = 4\lambda f^2/R^2$ specifying the diffraction spot-size along the optical axis for a lens with focus f. In this case, the multifocus lens of Eqs (6.58) and (6.71) does not produce separate foci but creates a continuous intensity distribution along the optical axis. This effect is similar to one when a single period of the diffraction grating does not secure the disintegration of a focal pattern into the set of focal spots.

As an example, we have computed the intensity distribution along the

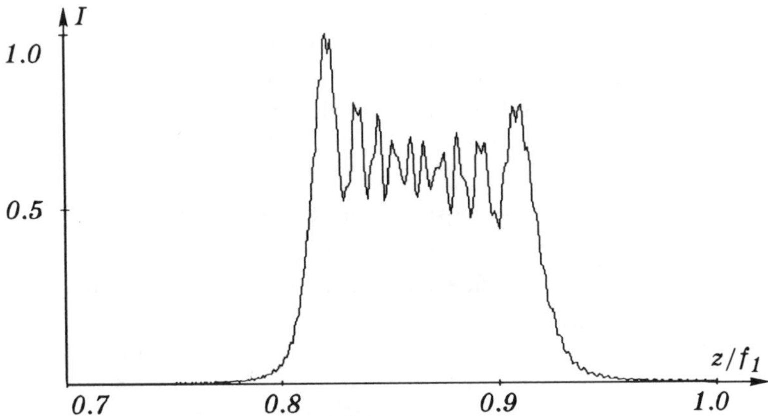

Figure 6.15 Axial intensity distribution for the multifocus lens of Eq. (6.70) at $f_2 = R^2/2\lambda$

optical axis of the lens defined in Eqs (6.70) and (6.71) with the following parameters: $N = 40$ ($n_1 = 20$, $n_2 = 60$), $\lambda = 1.06\,\mu$m, $R = 4$ mm, $f_1 = 30$ mm, and $f_2 = R^2/2\lambda = 7547.17$ mm. The result of the computation (see Fig. 6.15) demonstrates a continuous intensity distribution along the optical axis.

6.4 Computation of Two-order DOEs

Multifocus focusators dealt with in sections 6.1 and 6.2 allow one to realize focusing into a set of focal lines of different scale but equal form. In the present section, we discuss a modification of the earlier approach to synthesizing 'two-order DOEs' aimed at focusing into two different focal lines. Let us consider the procedure of calculating two-order DOEs. The domain of focusing is assumed to consist of two focal lines L_1 and L_2. The starting data for computing a two-order DOE are the phase functions $\varphi_a(\mathbf{u})$ and $\varphi_b(\mathbf{u})$ that provide focusing of the illuminating beam into the focal lines L_1 and L_2, respectively.

We propose that the phase function of the two-order DOE should be defined in the form [143]

$$\varphi(\mathbf{u}) = \tfrac{1}{2}[\varphi_a(\mathbf{u}) + \varphi_b(\mathbf{u})] + \Phi[h(\mathbf{u})] \tag{6.73}$$

where

$$h(\mathbf{u}) = \mathrm{mod}_{2\pi}\{\tfrac{1}{2}[\varphi_a(\mathbf{u}) - \varphi_b(\mathbf{u})]\}$$

As before, the function $\Phi[h(\mathbf{u})]$ of Eq. (6.73) describes a specially fitted nonlinear predistortion of the function $h(\mathbf{u})$.

To determine the form of the function $\Phi[\xi]$, let us analyze an operation of the element specified by Eq. (6.73). By expanding into a Fourier series the function $\exp(i\Phi[\xi])$ in the interval $[0, 2\pi]$, putting $\xi = h(\mathbf{u})$, and accounting for the 2π-periodicity, the complex transmittance of the element in Eq. (6.73) can be rewritten as

$$\exp[i\varphi(\mathbf{u})] = \sum_{j=-\infty}^{\infty} c_j \exp\left[\frac{i}{2}[\varphi_a(\mathbf{u})(1+j) + \varphi_b(\mathbf{u})(1-j)]\right] \tag{6.74}$$

where c_j are the Fourier coefficients of Eq. (6.19) in the decomposition of the function $\exp(i\Phi[\xi])$.

According to Eq. (6.74), an image formed in the jth diffraction order of the two-order DOE corresponds to the transformation of the illuminating beam by the element with the phase function given by

$$\varphi_j(\mathbf{u}) = \frac{1}{2}[\varphi_a(\mathbf{u})(1+j) + \varphi_b(\mathbf{u})(1-j)] \tag{6.75}$$

Focusing into the desired focal lines L_1 and L_2 occurs in the 1st and -1st diffraction orders; $\varphi_1(\mathbf{u}) = \varphi_a(\mathbf{u})$, and $\varphi_{-1}(\mathbf{u}) = \varphi_b(\mathbf{u})$. The portion of illuminating beam energy focused into the jth order is proportional to the squared modulus of the Fourier coefficients c_j. Therefore, the nonlinear predistortion function $\Phi[\xi]$ has to be chosen from the condition for the Fourier coefficients c_j to be zero at $j \neq \pm 1$. Such a function $\Phi[\xi]$ corresponds to the phase function of the diffraction grating concentrating the radiation into the 1st and -1st diffraction orders. To form a uniform intensity along the focal lines L_1 and L_2, the values of $|c_1|^2$ and $|c_{-1}|^2$ are chosen to be proportional to the geometrical size of the lines L_1 and L_2. When $|c_1|^2 \neq |c_{-1}|^2$, the calculation of $\Phi[\xi]$ is conducted using iterative methods. When $|c_1|^2 = |c_{-1}|^2$, the function $\Phi[\xi]$ can be determined as the phase function of the two-order binary grating of Eqs (6.44) and (6.45). For the grating of Eq. (6.44), we have $|c_1|^2 = |c_{-1}|^2 = 0.405$. In that case, a two-order DOE of Eqs (6.73) and (6.44) focuses 81% of the illuminating beam energy into the 1st and -1st diffraction orders.

No phase two-order grating with a 100% energy efficiency exists. Therefore, along with the desired focal lines L_1 and L_2, the two-order DOE of Eq. (6.73) also generates spurious images corresponding to the non-zero coefficients c_j at $j \neq \pm 1$. The degree of influence of the spurious images is estimated in each particular task and, if necessary, can be reduced via 'off-axial' focusing. This involves the introduction of prism phase functions into the functions $\varphi_a(\mathbf{u})$ and $\varphi_b(\mathbf{u})$. The prisms play the part of the carrier and make it possible to shift spurious images relative to useful ones.

The discussed method of computing two-order DOEs imposes no limits on the form of the functions $\varphi_a(\mathbf{u})$ and $\varphi_b(\mathbf{u})$. For example, if the functions $\varphi_a(\mathbf{u})$ and $\varphi_b(\mathbf{u})$ in Eq. (6.73) are specified as the phase functions of multifocus focusators for generating the sets of lines S_1 and S_2, a two-order

DOE generates in the 1st and −1st diffraction orders the sets of focal lines S_1 and S_2. In the examples which follow we point out the most practical cases of focusing into two points and two line segments of different direction.

As a simple example illustrating the method for designing two-order DOEs, we consider the calculation of a 'two-order lens' focusing the plane beam into two axial points at $z = f_a$ and $z = f_b$. The phase function of a two-order lens takes the form of Eq. (6.73), where $\Phi[\xi]$ is the phase function of the two-order grating of Eq. (6.44), the functions $\varphi_a(\mathbf{u})$ and $\varphi_b(\mathbf{u})$ being the phase functions of lenses with foci f_a and f_b. In a non-paraxial approximation, $\varphi_a(\mathbf{u})$ and $\varphi_b(\mathbf{u})$ are given by

$$\begin{cases} \varphi_a(\mathbf{u}) = -k\sqrt{(f_a^2 + \mathbf{u}^2)} \\ \varphi_b(\mathbf{u}) = -k\sqrt{(f_b^2 + \mathbf{u}^2)} \end{cases} \tag{6.76}$$

Contrary to the bifocal lens of Eqs (6.58), (6.67) and (6.44) discussed in section 6.3, the two-order lens of Eqs (6.73), (6.44), and (6.76) is a generalization of the case of non-paraxial approximation.

As another example, we consider the calculation of a two-order DOE focusing into a cross consisting of two perpendicular segments of length $2d$. In this case, the function $\Phi[\xi]$ also takes the form of Eq. (6.44), whereas the functions $\varphi_a(\mathbf{u})$ and $\varphi_b(\mathbf{u})$ in Eq. (6.73) are the phase functions of DOEs focusing into the line segments forming the cross. The functions $\varphi_a(\mathbf{u})$ and $\varphi_b(\mathbf{u})$ can be derived via iterative methods or using analytical Eqs (6.6) and (6.7) for the phase function of a focusator focusing into a line segment. Given the square aperture $G(2a \times 2a)$ and plane illuminating beam, analytical expressions for the functions $\varphi_a(\mathbf{u})$ and $\varphi_b(\mathbf{u})$ are easy to obtain from Eqs (6.6) and (6.7) in the form

$$\begin{cases} \varphi_a(\mathbf{u}) = -\dfrac{k\mathbf{u}^2}{2f} + \dfrac{kd}{2fa} u^2 \\ \varphi_b(\mathbf{u}) = -\dfrac{k\mathbf{u}^2}{2f} + \dfrac{kd}{2fa} v^2 \end{cases} \tag{6.77}$$

where f is the distance between the two-order DOE aperture and the focal plane.

Let us analyze the spurious images emerging in the course of focusing into a cross. According to Eq. (6.75), the structure of spurious images is determined by the phase functions

$$\varphi_j(\mathbf{u}) = -\frac{k\mathbf{u}^2}{2f} + \frac{kd}{4fa}[(1 + j)u^2 + (1 - j)v^2] \tag{6.78}$$

Equation (6.78) corresponds to the phase function of a focusator of a plane beam focusing into a rectangle [143] with sides $|d(1 + j)|$ and $|d(j - 1)|$. Due to the uniform energy distribution over the rectangle area, the intensity of spurious orders will be significantly less than the intensity on the focal lines

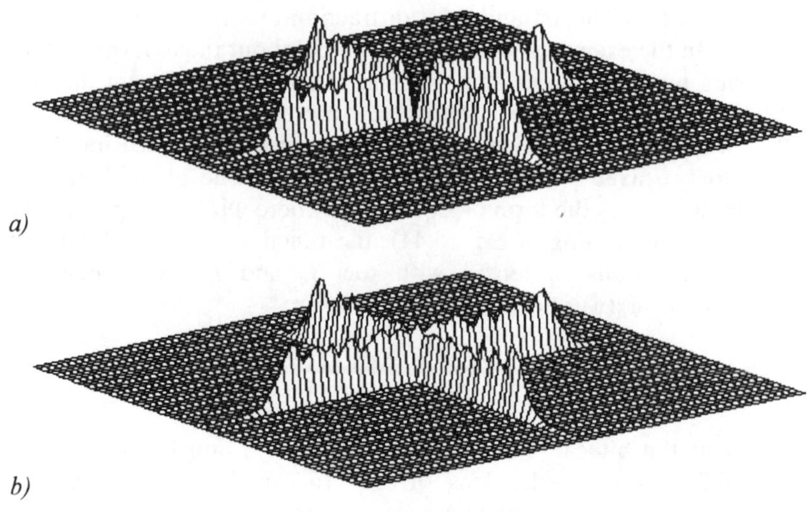

Figure 6.16 Intensity distribution in the focal plane of two-order DOEs focusing into a cross: (a) for the function $\Phi[\xi]$ of Eq. (6.44); (b) upon the cyclic shift of profile $\Phi[\xi]$ by $\pi/6$

making up the cross. This implies that spurious orders should not significantly affect the quality of the image produced.

With the objective of analyzing the developed approach, we have conducted the diffractive calculation of the intensity in the focal plane of the two-order DOE of Eqs (6.73), (6.44), and (6.77) using the following parameters: $\lambda = 1.06\,\mu$m, $f = 100$ mm, $2d = 1.2$ mm, and $2a = 5$ mm (see Fig. 6.16a). The structure of the crosswise focused radiation is clearly seen in Fig. 6.16a. Spurious images are not seen, but significant intensity fall-off is found in the cross centre. The fall-off is due to different signs of the Fourier coefficients c_1 and c_{-1} in Eq. (6.45). According to Eq. (6.74), this leads to the subtraction of the fields corresponding to focusing into the cross-segments. The intensity fall-off in the cross centre can be eliminated by means of the cyclic shift of profile of the grating in Eq. (6.44). If the profile of the grating in Eq. (6.44) undergoes a cyclic shift by the value of φ_0, the Fourier coefficients of Eq. (6.45) will take the form

$$c_j^{\varphi_0} = c_j \exp(ij\varphi_0) \tag{6.79}$$

The field amplitude in the cross centre is proportional to the quantity $\eta = c_1 \exp(i\varphi_0) + c_{-1}\exp(-i\varphi_0) = 2ic_1 \sin(\varphi_0)$. When $\varphi_0 = \pi/6$, $\eta = ic_1$ and the intensity dip in the cross centre should disappear. The calculated distribution of the field intensity in the focal plane of the two-order DOE (see Fig. 6.16b) obtained via the cyclic shift by $\pi/6$ of the profile of the grating in Eq. (6.44) shows the absence of the intensity fall-off in the cross centre.

7

Calculation of DOEs for Some Special Applications

7.1 Focusing into a Transverse Line Segment

The problem of focusing into a line segment can be treated as a reference problem of the generation of a complex focal line. The calculation of DOEs focusing into a line considered in Chapter 6 is feasible only for simple smooth lines such as a line segment, a ring, a semiring, or an arc of parabola [132,133,136,144]. The complex focal line may be approximated by a set of line segments with some practically feasible accuracy. To focus into the set of line segments, one can employ segmentized DOEs. In that case, the DOE's aperture is broken down into segments (in accordance with the number of focal segments), each of them focusing into a corresponding part of the focal line under synthesis [133,144]. In the present section, an analytical and numerical analysis of the solutions to the problem of focusing into a segment is given, allowing the type of DOE phase function to be chosen as required by the particular technological task.

7.1.1 *Analytical Calculation of a DOE Focusing into a Line Segment*

The method for computing DOEs focusing into a line that is treated in Chapter 6 is based on a ray-tracing approach. Since the focal line is a caustic one, diffractive effects are significant upon focusing into a line. When calculating focusators, the diffraction blurring of the focal line is assumed to be zero; the function $\theta(x)$ in Eq. (6.7) characterizes the energy distribution of focused light along the focal line without regard for a real diffractive width [134,135].

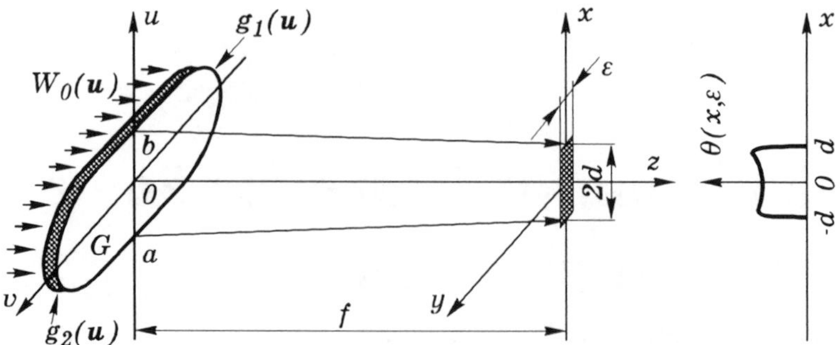

Figure 7.1 Geometry of the problem of focusing into a line segment

Herein, we discuss an analytical method for calculating DOEs intended for focusing into a line segment, taking account of the diffraction blurring of the focal line.

Let us consider the calculation of a DOE focusing the laser light with amplitude $W_0(\mathbf{u}) = \sqrt{[I_0(\mathbf{u})]}\exp[i\varphi_0(\mathbf{u})]$, where $I_0(\mathbf{u})$ is the illuminating beam intensity and $\varphi_0(\mathbf{u})$ is the beam phase, into a segment $|x| \le d$ in the plane $z = f$ (see Fig. 7.1). The DOE aperture is assumed to be confined by the curves $v = g_1(u)$ and $v = g_2(u)$, and by the line segments $u = a$ and $u = b$. It is desired to find the DOE phase function $\varphi(\mathbf{u})$ providing the generation in the focal plane $z = f$ of the light field with the intensity distribution $I(\mathbf{x})$ obeying the condition

$$\int_{-\varepsilon/2}^{\varepsilon/2} I(x, y)\,\mathrm{d}y = \theta(x, \varepsilon), \quad |x| \le d \tag{7.1}$$

The function $\theta(x, \varepsilon)$ characterizes the energy distribution in an ε-vicinity of the focal segment. When $\varepsilon \ll \Delta$ (Δ is the segment diffraction width), the $\theta(x, \varepsilon)$ function is proportional to the value of intensity $I(x, 0)$ over the geometrical segment.

We shall seek a phase function $\varphi(\mathbf{u})$ such that all the rays drawn from an arbitrary 'focusing curve' $u = c$ intersect the plane of focusing in one point $(x(u), 0)$ of the focal segment. We assume that the eikonal equation holds in a focusator vicinity. Then, the ray-tracing equations give us the phase function of the DOE in a paraxial approximation, in the form [133–135]

$$\varphi(\mathbf{u}) = -\frac{k\mathbf{u}^2}{2f} + \frac{k}{f}\int_{u_0}^{u} \chi(\xi)\,\mathrm{d}\xi - \varphi_0(\mathbf{u}) \tag{7.2}$$

where $k = 2\pi/\lambda$ and λ is the wavelength.

To derive the $\chi(u)$ function in Eq. (7.2), let us obtain the diffractive approximation for the field in the vicinity of the focal segment. Substitution of Eq. (7.2) into the Fresnel–Kirchhoff integral gives us the field complex amplitude in the DOE focal plane in the form

$$F(\mathbf{x}) = \frac{1}{\lambda i f} \exp\left(i \frac{k\mathbf{x}^2}{2f} \right) \int_G \sqrt{[I_0(\mathbf{u})]} \times \exp\left[i \frac{k}{f} \int_{u_0}^{u} \chi(\xi)\,\mathrm{d}\xi - i \frac{k}{f} \mathbf{xu} \right] \mathrm{d}^2\mathbf{u} \tag{7.3}$$

Using the stationary phase method while integrating over the focusing curve [145] (i.e. with respect to u) in Eq. (7.3) yields the following diffraction approximation for the intensity distribution $I(\mathbf{x}) = |F(\mathbf{x})|^2$ [133–135]

$$I[\chi(u), y] = \left| \sqrt{\left(\frac{1}{\lambda f} \right)} \int_{g_1(u)}^{g_2(u)} \sqrt{[I_0(u, v)]} \left(\frac{\mathrm{d}\chi(u)}{\mathrm{d}u} \right)^{-1/2} \exp\left(-i \frac{k}{f} yv \right) \mathrm{d}v \right|^2 \tag{7.4}$$

The approximation of Eq. (7.4) describes diffraction effects within the focal segment, but not near its ends. Substituting Eq. (7.4) into Eq. (7.1), we obtain

$$\theta[\chi(u), \varepsilon] = \int_{-\varepsilon/2}^{\varepsilon/2} \left| \sqrt{\left(\frac{1}{\lambda f} \right)} \int_{g_1(u)}^{g_2(u)} \sqrt{[I_0(u, v)]} \left(\frac{\mathrm{d}\chi(u)}{\mathrm{d}u} \right)^{-1/2} \exp\left(-i \frac{k}{f} yv \right) \mathrm{d}v \right|^2 \mathrm{d}y \tag{7.5}$$

According to Eq. (7.5), the function $\chi(u)$ in Eq. (7.2) must be found as a solution of the following differential equation [133–136]

$$\frac{\mathrm{d}\chi(u)}{\mathrm{d}u} \theta(\chi(u), \varepsilon) = \int_{-\varepsilon/2}^{\varepsilon/2} \left| \sqrt{\left(\frac{1}{\lambda f} \right)} \int_{g_1(u)}^{g_2(u)} \sqrt{[I_0(u, v)]} \exp\left(-i \frac{k}{f} yv \right) \mathrm{d}v \right|^2 \mathrm{d}y \tag{7.6}$$

with boundary conditions $\chi(a) = -d, \chi(b) = d$.

Let us now consider some special cases of the DOE specified by Eqs (7.2) and (7.6). When the illuminating beam is uniform ($I_0(\mathbf{u}) = I_0$), Eq. (7.6) takes the form

$$\frac{\mathrm{d}\chi(u)}{\mathrm{d}u} \theta[\chi(u), \varepsilon] = c\Phi\left(\frac{k\varepsilon}{4f} [g_2(u) - g_1(u)] \right) [g_2(u) - g_1(u)] \tag{7.7}$$

where c is a constant

$$\Phi(\beta) = \mathrm{Si}(2\beta) - \beta\,\mathrm{sinc}^2(\beta)$$

$$\mathrm{Si}(\beta) = \int_0^{\beta} \frac{\sin(x)}{x} \mathrm{d}x$$

$$\mathrm{sinc}(\beta) = \frac{\sin(\beta)}{\beta}$$

When $\varepsilon \ll \Delta$, where Δ is the segment diffraction width, we will expand both sides of Eq. (7.6) in ε degrees and consider only linear members of the expansion. The resulting equation for the $\chi(u)$ function is

$$\frac{\mathrm{d}\chi(u)}{\mathrm{d}u} I[\chi(u)] = \left| \sqrt{\left(\frac{1}{\lambda f}\right)} \int_{g_1(u)}^{g_2(u)} \sqrt{[I_0(u,v)]}\,\mathrm{d}v \right|^2 \tag{7.8}$$

$$\chi(a) = -d, \chi(b) = d$$

where $I(x)$ is a predetermined intensity along the segment at $y = 0$ and $|x| \leq d$.

When $\varepsilon \gg \Delta$, let us replace the external definite integral in Eq. (7.5) by an indefinite integral and apply the Parseval equality. As a result, Eq. (7.6) takes the form

$$\begin{gathered} \frac{\mathrm{d}\chi(u)}{\mathrm{d}u} \theta[\chi(u)] = \int_{g_1(u)}^{g_2(u)} I_0(u,\xi)\,\mathrm{d}\xi \\ \chi(a) = -d, \chi(b) = d \end{gathered} \tag{7.9}$$

where

$$\theta(x) = \theta(x,\varepsilon)|_{\varepsilon \gg \Delta} = \int_{-\infty}^{\infty} I(x,y)\,\mathrm{d}y$$

It should be noted that Eqs (7.2) and (7.9) coincide with the known representation of the focusator phase function [37,126,145]. The focusators are limited by the condition $\varepsilon \gg \Delta$. The DOEs of Eqs (7.2), (7.6)–(7.8) provide the formation of the predetermined energy distribution $\theta(x, \varepsilon)$ along the focal segment for arbitrary ε. We shall call the DOE with the phase function of Eqs (7.2) and (7.6)–(7.8) the focusator with diffraction corrections (FDC).

Let us make a comparative analysis of the efficiency of the focusator specified by Eqs (7.2) and (7.9), and the FDC of Eqs (7.2) and (7.6). For estimating the DOE performance, we will use the energy efficiency E and the r.m.s. deviation δ. The quantity

$$E(\varepsilon) = \int_{-d}^{d} \theta(x,\varepsilon)\,\mathrm{d}x \left[\int_G I_0(\mathbf{u})\,\mathrm{d}^2\mathbf{u} \right]^{-1} \tag{7.10}$$

characterizes the portion of the illuminating beam energy focused into an ε-vicinity of the focal segment.

The quantity

$$\delta(\varepsilon) = \frac{1}{\bar{\theta}} \left[\frac{1}{2d} \int_{-d}^{d} [\theta(x,\varepsilon) - \bar{\theta}]^2\,\mathrm{d}x \right]^{1/2} \tag{7.11}$$

Table 7.1 Values of E and δ for a focusator and for FDCs

	Focusator		FDCs	
ε/Δ	E (%)	δ (%)	E (%)	δ (%)
$\varepsilon \ll \Delta$	—	28.7	—	13.3
0.25	62.3	36.3	62.2	16.2
0.50	83.7	18.6	83.6	14.8
0.75	89.7	18.2	89.5	15.0
1.00	90.5	15.6	90.8	15.2

characterizes the r.m.s. deviation of the energy distribution $\theta(x, \varepsilon)$ from the mean value

$$\bar{\theta} = \frac{1}{2d} \int_{-d}^{d} \theta(x, \varepsilon)\,\mathrm{d}x$$

For $\varepsilon \ll \Delta$, $\delta(\varepsilon)$ corresponds to the r.m.s. deviation of the intensity distribution along the geometrical segment from a constant value.

Table 7.1 gives the values of $E(\varepsilon)$ and $\delta(\varepsilon)$ for a focusator intended for focusing a uniform ring-section beam (radii R_1 and R_2) into a line segment with a uniform energy distribution for the following parameters: $\lambda = 1.06\ \mu$m, $f = 100$ mm, $2d = 1$ mm, $R_1 = 3$ mm, and $R_2 = 5$ mm. It also gives analogous data for FDCs. The quantity Δ is the diffractive width at the centre of the focal segment: $\Delta = 2\lambda f/(R_2 - R_1)$. According to Table 7.1, for the focusator of Eqs (7.2), and (7.9), the energy distribution appears to be most nonuniform for $\varepsilon \ll \Delta$ and $\varepsilon = \Delta/4$; $\delta(\varepsilon)|_{\varepsilon \ll \Delta} = 28.7\%$ and $\delta(\Delta/4) = 36.3\%$ (see Figs 7.2a and 7.2b). At the above specified values of ε, FDCs give an r.m.s. deviation $\delta(\varepsilon)$ which is less than half that for the focusator (see Figs 7.3a and 7.3b). When $\varepsilon = \Delta/2$, the FDC allows the reduction of $\delta(\varepsilon)$ only by a factor of 1.26 (see Figs 7.2c and 7.3c), and when $\varepsilon = \Delta$, the focusator and the FDC offer practically identical performances. As is seen from the pursued studies, the FDCs allow the formation of a desired energy distribution $\theta(x, \varepsilon)$ along the focal segment for an arbitrary ε with a 13–16% error. The focusator represented by Eqs (7.2) and (7.9) is a particular case of the FDC, which is intended for the production of a desired energy distribution at $\varepsilon \geq \Delta$.

7.1.2 Comparative Analysis of Iterative and Analytical Methods

It appears to be topical to make a comparative analysis of the DOEs calculated by the iterative Gerchberg–Saxton (GS) algorithm and FDCs. Such an analysis will be performed for the case of forming a predetermined intensity distribution $I(x)$ along the focal segment.

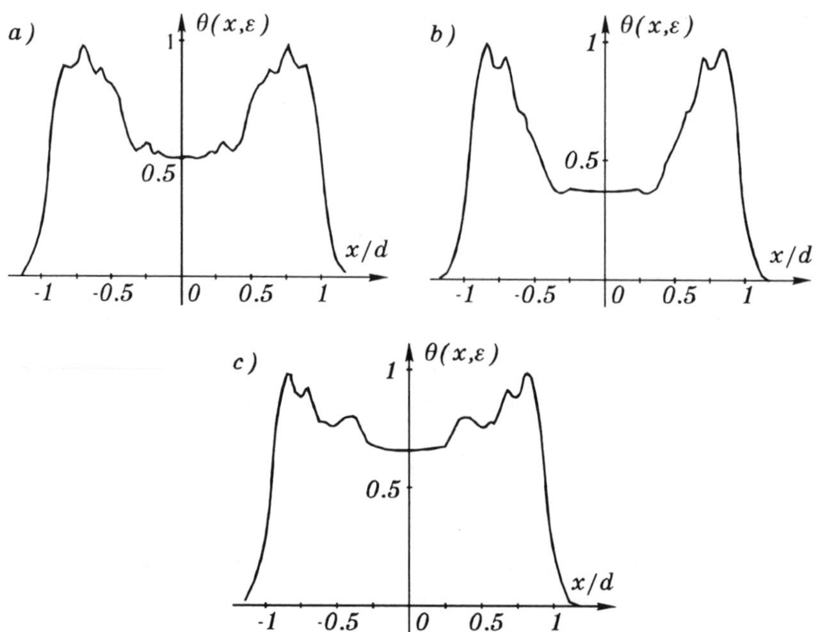

Figure 7.2 Function $\theta(x, \varepsilon)$ for a focusator: (a) $\varepsilon \ll \Delta$; (b) $\varepsilon = \Delta/4$; (c) $\varepsilon = \Delta/2$

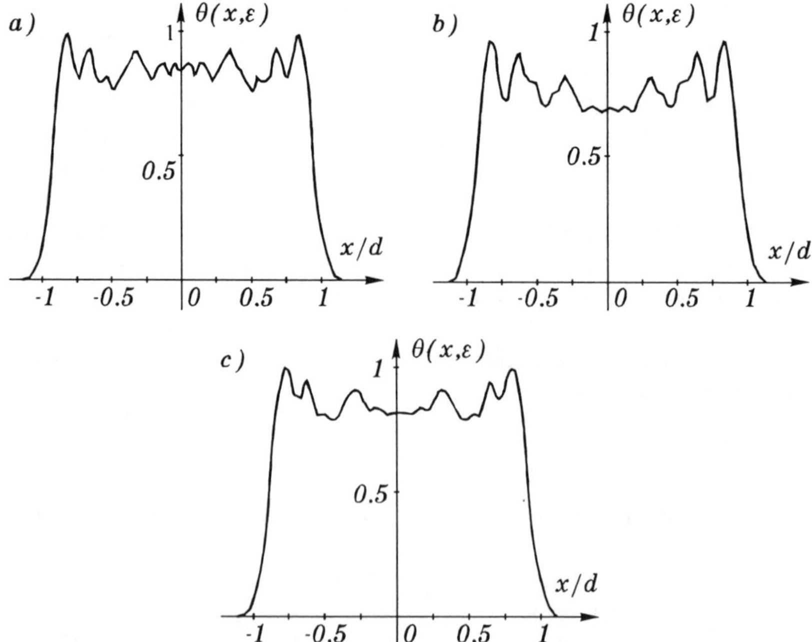

Figure 7.3 Function $\theta(x, \varepsilon)$ for FDCs: (a) $\varepsilon \ll \Delta$; (b) $\varepsilon = \Delta/4$; (c) $\varepsilon = \Delta/2$

The phase function of the FDC has the form

$$\varphi(\mathbf{u}) = -\frac{k\mathbf{u}^2}{2f} + f(u) - \varphi_0(\mathbf{u}) \tag{7.12}$$

where

$$f(u) = \frac{k}{f}\int_{u_0}^{u} \chi(\xi)\,\mathrm{d}\xi \tag{7.13}$$

The above calculation of the $f(u)$ function was based on the analytical choice of a smooth correspondence $x = \chi(u)$ between the focusing curves and the points of the focal segment. Let us consider the calculation of the function $f(u)$ using the iterative GS algorithm. The field complex amplitude formed by the DOE with the phase function given in Eq. (7.12) has the form

$$F(x, y) = \frac{1}{\lambda i f}\exp\left(i\frac{k\mathbf{x}^2}{2f}\right)\int_a^b \gamma(u, y)\exp\left(if(u) - i\frac{k}{f}xu\right)\mathrm{d}u \tag{7.14}$$

where

$$\gamma(u, y) = \int_{g_1(u)}^{g_2(u)} \sqrt{[I_0(u, v)]}\exp\left(-i\frac{k}{f}yv\right)\mathrm{d}v \tag{7.15}$$

Equations (7.14) and (7.15) were obtained from the Fresnel–Kirchhoff integral taken over a variable v. According to Eq. (7.14), the intensity distribution $I(x) = |F(x, 0)|^2$ along the x-axis (i.e. along the focal segment) corresponds to the intensity distribution formed by a 1D DOE, with the phase function $f(u)$ illuminated by a beam with complex amplitude $\gamma(u, 0)$ $\exp[-i(ku^2/2f)]$. Therefore, for the formation of the desired intensity distribution along the focal segment we can calculate the function $f(u)$ using the 1D iterative algorithms, thus essentially reducing computational efforts.

To estimate the DOE performance, we will use the energy efficiency $E(\Delta)$ (see Eq. (7.10)) and the r.m.s. deviation δ of the intensity distribution $I(x)$ from the mean value $\bar{I}$.

Table 7.2 gives the values of $E(\Delta)$ and δ versus the parameter $\eta = 2d/\Delta$ characterizing the focal segment length in comparison with diffraction width Δ in the centre of the focal segment. The values presented in Table 7.2 correspond to focusing a uniform circle beam with radius R into a segment with uniform intensity distribution ($\Delta = \lambda f/R$). Table 7.2 gives the $E(\Delta)$ and δ values for the FDC in Eqs (7.2) and (7.8), and for the DOEs calculated using a GS algorithm. Note that the values are shown with random initial phase, and also with the phase function of the FDC of Eqs (7.2) and (7.8) taken as an initial phase for the GS algorithm. In the last two cases, the calculation was conducted via a fast Fourier transform for the following parameters: the number of pixels is $N = 512$ and the number of pixels within the DOE aperture is $N_u = 256$. In this case, the focal plane step is $\Delta x = \Delta/4$,

Table 7.2 Values of $E(\Delta)$ and δ for FDCs and for DOEs calculated via a GS algorithm

			Iterative elements			
	FDC		Starting-point – random phase		Starting-point – phase function of FDC	
$2d/\Delta$	E (%)	δ (%)	E (%)	δ (%)	E (%)	δ (%)
4	79.9	26.7	81.3	31.7	83.4	26.8
8	82.3	21.3	82.9	28.7	84.9	17.1
16	84.0	16.6	83.4	28.4	85.6	12.6
32	84.8	14.6	82.7	33.0	85.9	13.3

and the number of pixels within the focal segment with length $2d = \eta\Delta$ corresponds to the value of $N_x = 4\eta$. Table 7.2 shows that the FDC allows one to form more uniform intensity distributions along the focal segment in comparison with 'iterative' DOEs calculated for the random initial phase. If the phase function $f(u)$ of an FDC is used as the starting-point for the GS algorithm, the r.m.s. deviation decreases by 1–4% and the energy efficiency increases by 1–3%.

Chapter 1 deals with an adaptive–additive (AA) algorithm allowing the enhancement of the iterative process convergence. Table 7.3 gives the values of $E(\Delta)$ and δ for the DOEs calculated using the AA algorithm. As can be seen from Table 7.3, the DOE calculated using the AA algorithm, with the phase function of an FDC taken as the initial phase, offers an r.m.s. deviation δ which is 3–7 times smaller than that for the FDC. Note that the values of energy efficiency differ insignificantly. The application of the AA algorithm for the random initial phase results in a 3–15% decrease of δ and in a severe decrease of the energy efficiency by 8–15%.

Table 7.3 Values of $E(\Delta)$ and δ for DOEs calculated via an AA algorithm

	Starting-point			
	Random phase		Phase function of FDC	
$2d/\Delta$	E (%)	δ (%)	E (%)	δ (%)
4	72.0	10.5	76.5	3.8
8	71.1	9.9	81.6	5.5
16	69.5	13.4	83.8	5.1
32	69.3	14.5	83.7	4.3

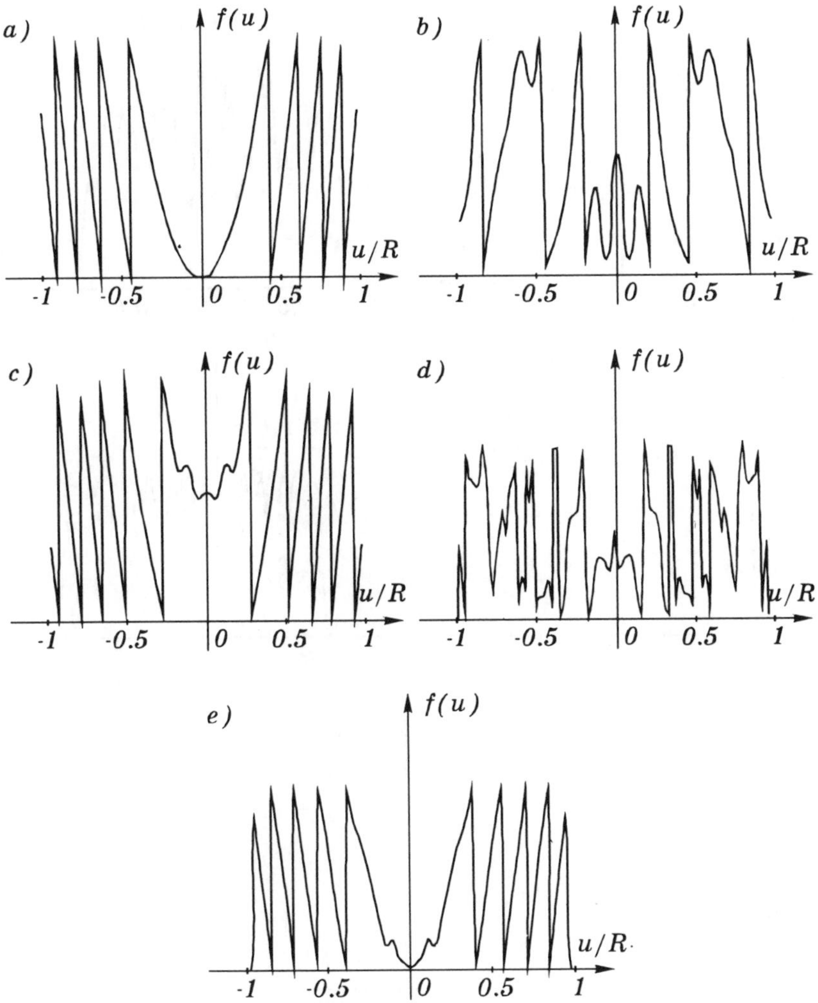

Figure 7.4 Phase functions of DOEs focusing into a line segment of length 16Δ: (a) phase of an FDC; (b), (c) phases of DOEs calculated via a GS algorithm; (d), (e) phases of DOEs calculated via an AA algorithm

By way of illustration, Fig. 7.4 shows the $f(u)$ functions calculated for the following cases: for the FDC; for the DOEs calculated via the GS algorithm with a random initial phase and with the phase function of an FDC taken as the initial phase; and for the DOEs calculated using the AA algorithm with random initial phase and with the phase function of an FDC taken as the starting phase. These plots have been obtained for focusing into a line segment of length $2d = 16\Delta$. Figure 7.5 depicts calculated intensity distributions along the focal segments for the $f(u)$ functions shown in Fig. 7.4.

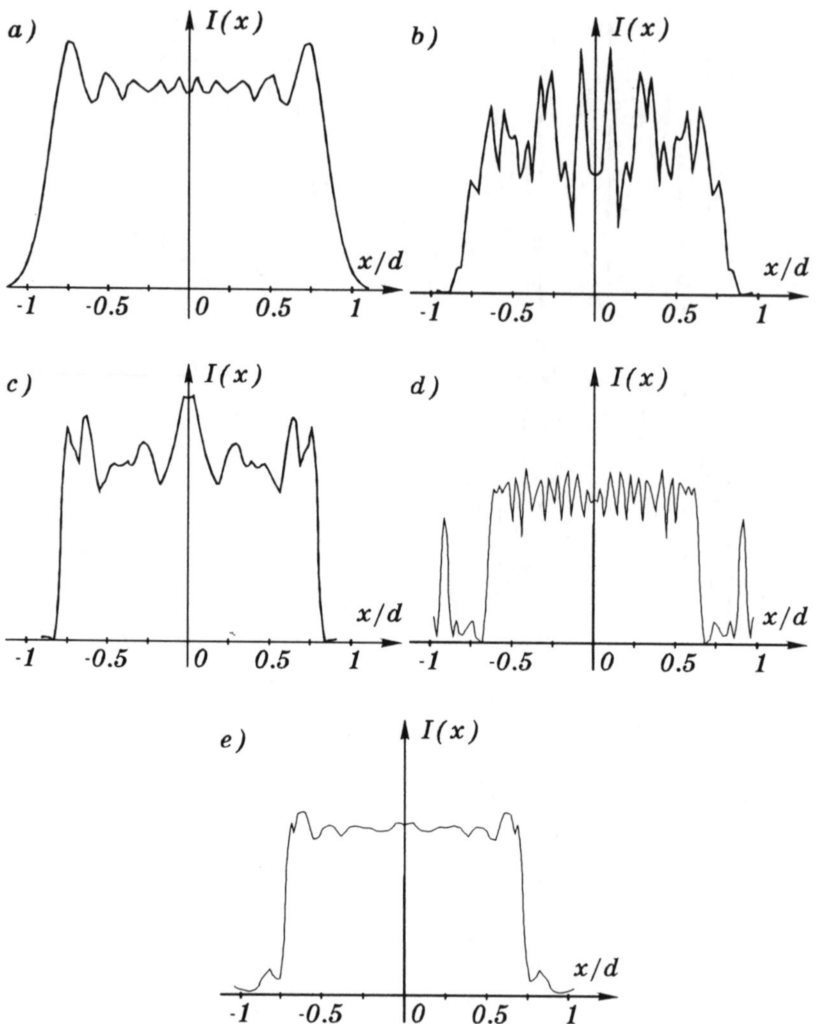

Figure 7.5 Intensity distributions along the focal segments for the phase functions represented in Fig. 7.4

Intensity peaks at $|x| \geq d$ in Fig. 7.5d are due to the fact that the AA algorithm disregards the form of intensity distribution exterior to the focal segment. At the same time, the AA algorithm introduces no deterioration into the intensity distribution outside the focal segment if the phase function of the FDC is used as the starting-point for the AA algorithm (see Fig. 7.5e).

Numerical research has shown that in comparison with the DOEs derived via the GS algorithm with random initial guess (see Table 7.2), the FDC enables the formation of a more uniform intensity distribution along the focal segment. The use of the FDC as the starting approximation for the GS algorithm can offer only insignificant enhancement of the quality of focusing, while the employment of the AA algorithm in combination with the phase function of the FDC in place of the starting approximation results in a 3–5-fold decrease in the non-uniformity of the intensity distribution along the focal segment in comparison with the FDC (see Tables 7.2 and 7.3).

7.2 Focusing of Gaussian Beam into a Square

The demand for optical elements capable of making the Gaussian intensity distribution uniform exists in such areas of optical data processing as laser-based superficial hardening and laser printing.

It has been shown [146,147] that for a DOE that forms the rectangular intensity-uniform distribution given by

$$I(x,y)=\begin{cases}1, & |x|\leq d_1, \quad |y|\leq d_2\\ 0, & |x|>d_1, \quad |y|>d_2\end{cases}$$

from the plane beam with the Gaussian profile of intensity distribution

$$I_0(u,v)=I_0\exp\left[-\frac{u^2+v^2}{\sigma^2}\right]$$

one can (given a lens located immediately adjacent to the DOE) derive its phase function in the form of a sum

$$\Phi(u,v)=\Phi_1(u)+\Phi_2(v)$$

with its terms satisfying the set of two simultaneous equations

$$\begin{cases}\dfrac{\mathrm{d}x}{\mathrm{d}u}=\dfrac{I_0(u)}{I(x)}\\ x=fk^{-1}\dfrac{\mathrm{d}\Phi(u)}{\mathrm{d}u}\end{cases} \tag{7.16}$$

where f is the focal length of the lens in whose focal plane the light rectangle is formed, k is the wavenumber of light, (u,v) and (x,y) are the coordinates in the planes of the DOE and Fourier spectrum, respectively, $2d_1$ and $2d_2$ are the measures of the rectangle, and σ is the parameter of the Gaussian beam.

The first equation in the system (7.16) sets the equality of the density of the light energy of a 1D DOE to the density of the corresponding areas of the straight-light segment. The second equation in the system (7.16) describes the stationary points in the paraxial approximation.

The discrete variant of the solution of the system (7.16) can be written as

$$\Phi_{mn} = \ln 10 \left\{ M_x \left[\frac{6n}{N} \operatorname{erf}\left(\frac{6n}{N} \right) + \frac{1}{\sqrt{\pi}} e^{-36n^2/N^2} \right] \right.$$
$$\left. + M_y \left[\frac{6m}{N} \operatorname{erf}\left(\frac{6m}{N} \right) + \frac{1}{\sqrt{\pi}} e^{-36m^2/N^2} \right] \right\} \tag{7.17}$$

where

$$\operatorname{erf}(x) = \frac{2}{\sqrt{\pi}} \int_0^x e^{-t^2} dt$$

$$m = 0, \pm 1, \pm 2, \ldots, \pm \frac{N}{2}$$

$$n = 0, \pm 1, \pm 2, \ldots, \pm \frac{N}{2}$$

$$M_x = \frac{\pi N_x}{6\sqrt{(\ln 10)}}$$

$$M_y = \frac{\pi N_y}{6\sqrt{(\ln 10)}}$$

M_x and M_y are the numbers of the minimum diffraction spots that fall into the rectangle with measures $N_x \times N_y$. In this case, the square aperture of the DOE measuring $N \times N$ equals the side 6σ of the square, and the Gaussian collimated beam that illuminates the DOE produces the light distribution given by

$$I_{0_{mn}} = \exp[-36N^{-2}(n^2 + m^2)] \tag{7.18}$$

In further discussion we employ the phase of Eq. (7.17) to simulate numerically the operation of a ray optics DOE.

Following [26,27], let us briefly describe how iterative algorithms can apply to computing DOE as a kind of kinoform. We will assume the preset complex amplitude $A(u)$ of the illuminating beam and the required intensity distribution $I(x)$ in the focal plane of the lens. The complex amplitude $F(x)$ in the focal plane is related to the complex amplitude

$$f(u) = A(u) \exp[i\Phi(u)]$$

immediately behind the DOE through the Fourier transform

$$F(x) = \int_{-b}^{b} f(u) e^{-i(k/f)xu} dx$$

where $2b$ is the size of the DOE's aperture. The DOE's equation takes the form

$$|F(x)|^2 = I(x) \tag{7.19}$$

To find an iterative solution of Eq. (7.19) with respect to the phase $\Phi(u)$, one should perform some preliminary estimation of the phase $\Phi_0(u)$ followed by the computation of the complex light amplitude in the focal plane. In this case, the complex amplitude $F_n(x)$ calculated in the nth iteration step is replaced by the function $F_n^0(x)$ according to the rule

$$F_n^0(x) = \begin{cases} \sqrt{[I_n(x)]}\dfrac{F_n(x)}{|F_n(x)|}, & |x| \leq d \\ \sqrt{(\alpha)}F_n(x), & |x| > d \end{cases} \tag{7.20}$$

where $I_n(x) = (1+\alpha)I(x) - \alpha|F_n(x)|^2$, $I(x)$ is the required intensity distribution within the interval $[-d, d]$ of the focal plane, and α is the parameter that controls the rate of convergence of the calculated intensity to the required one. For $\alpha = 0$, the replacement in Eq. (7.20) changes to the standard replacement in the GS algorithm [14,148].

The amplitude of light in the plane of DOE $f_n(u)$ is calculated with the help of the inverse Fourier transform and is replaced by the function $f_n^0(x)$ according to the rule

$$f_n^0(x) = \begin{cases} A(u)\dfrac{f_n(u)}{|f_n(u)|}, & |u| \leq b \\ 0, & |u| > b \end{cases} \tag{7.21}$$

In contrast to the algorithms of the conditional gradient [149], the α parameter is introduced here directly into the intensity function.

The rate of convergence of the intensity $|F_n(x)|^2$ to the required one $I(x)$ is checked by the root-mean-square deviation

$$\delta = \left[\frac{\displaystyle\int_{-d}^{d} [I(x) - |F_n(x)|^2]^2 \mathrm{d}x}{\displaystyle\int_{-d}^{d} I^2(x)\,\mathrm{d}x} \right]^{1/2} \tag{7.22}$$

Also, a new parameter E characterizing the energy efficiency of focusing is introduced

$$E = \frac{\displaystyle\int_{-d}^{d} |F_n(x)|^2 \mathrm{d}x}{\displaystyle\int_{-\infty}^{\infty} |F_n(x)|^2 \mathrm{d}x} \tag{7.23}$$

In what follows, we apply the phase derived from Eqs (7.19)–(7.21) to numerical simulation of the operation of DOEs [150].

We examine the DOE in a square that comprises 32×32 pixels and measures 10 minimum diffraction spots. For a DOE that focuses from the Gaussian collimated beam into a square, the phase deduced from Eq. (7.17)

Figure 7.6 Ray optics phase of the DOE into a square

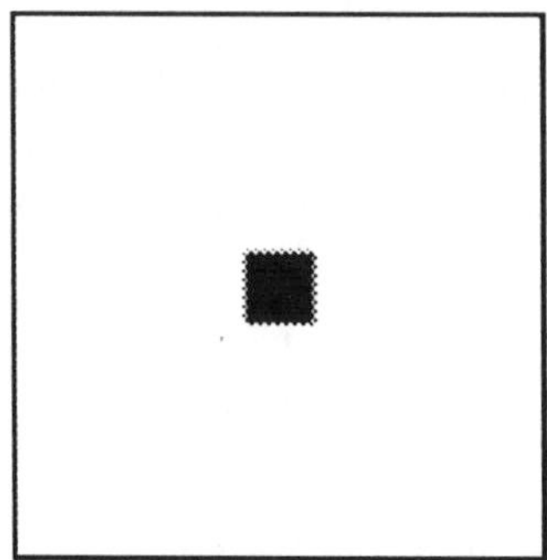

Figure 7.7 Intensity distribution obtained from the ray optics DOE in a lens focal plane

on the array of 256×256 pixels and taken to the modulus 2π is found as a set of rings (lines of equal phase) changing to the lines of square perimeter (Fig. 7.6). Figure 7.7 illustrates the light intensity distribution in the lens focal plane calculated as the Fourier transform of the amplitude

$$f_{mn} = \sqrt{(I_{0_{mn}})}\, e^{i\Phi_{mn}}$$

where $I_{0_{mn}}$ is taken from Eq. (7.18) and Φ_{mn} from Eq. (7.17). The root-mean-square deviation of the obtained distribution from the uniform one amounts to 5% and the efficiency is 91.6%.

To calculate the DOE that focuses the Gaussian beam into a uniform-intensity square, as an initial phase estimate, we have chosen the ray optics phase function described above (see Fig. 7.6). The subsequent iterative calculation has been conducted using three techniques. The first approach to the calculation of the phase based on the standard variant of the Gerchberg–Saxton algorithm with the replacement (7.20), for $\alpha = 0$, yields, at first, an increase of the error δ in the first iteration steps and then gives a slow decrease of the error (Fig. 7.8a, curve 1). In this case, during 10

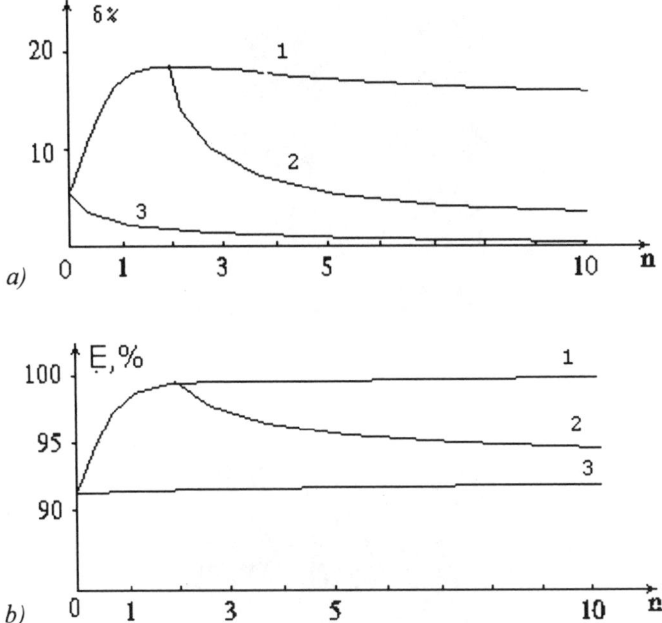

Figure 7.8 The root-mean-square deviation (a) and the energy efficiency (b) plotted against the number of iterations for different variants of an iterative algorithm: Gerchberg–Saxton (1), combined (2), and adaptive (3)

iterations the error does not diminish below 13% though the efficiency E increases to 98.9% (see Fig. 7.8b, curve 1).

The second approach is combined: the first three iterations are conducted with the replacement in Eq. (7.20) for $\alpha = 0$, the remaining seven iterations for $\alpha = 1$ (Fig. 7.8a, curve 2). This method after 10 iterations results in the formation of a square characterized by uniform intensity, with 2.4% error and 96.4% efficiency (Fig. 7.8b, curve 2). This way is seen to be more effective than the first approach, since it yields a six-fold decrease of the error from 13% to 2% without an essential decrease in the efficiency.

The third method is purely adaptive, which means that the replacement in Eq. (7.20), with $\alpha = 1$, is performed in each step of the iteration (Fig. 7.8a, curve 3). As one can see in this case, the error decreases monotonically and during 10 iterations it becomes equal to 0.1%, but the efficiency falls to 92.2% (Fig. 7.8b, curve 3). Figure 7.9 illustrates the DOE phase modulo 2π that has been calculated during 10 iterations, on the basis of the third approach, from the ray optics phase of Eq. (7.17). First, one can see that the use of the iterative algorithm does not result in an essential change of the initial phase. The pronounced changes of the phase (Fig. 7.6) take place only on the edges of the focusator (Fig. 7.9). Second, we can draw the conclusion that the adaptive iterative algorithm based on the replacement

Figure 7.9 The phase of the DOE into a square obtained after 10 iterations from the initial ray optics phase

Figure 7.10 The radius-random phase chosen as the initial estimate for iterations

of Eq. (7.20), for $\alpha = 1$, makes it possible to improve the ray optics phase in such a manner that the error in the formation of a light square reduces by more than an order, with the efficiency almost unchanged.

For comparison, we conducted another numerical simulation where we made use of a radius-random phase function (Fig. 7.10) as the initial approximation. The phase shown in Fig. 7.11 was calculated during 10 iterations. The intensity distribution formed by the DOE characterized by such a phase is presented in Fig. 7.12. The error and the efficiency were $\delta = 6.4\%$ and $E = 80.8\%$, respectively. Further iterations did not produce an essential change in these values. Data given above show that the iterative algorithm that begins with the random phase estimate leads to the non-regular structure of the DOE's zones and results in accuracy and efficiency that are somewhat lower than those for the ray optics DOE.

Table 7.4 summarizes all the DOEs discussed in this section. One can see the advantages in terms of uniformity, efficiency, and regular phase structure achieved using a physically justified initial approximation in the form of the

Figure 7.11 The phase of the DOE into a square obtained after 10 iterations from the initial random phase

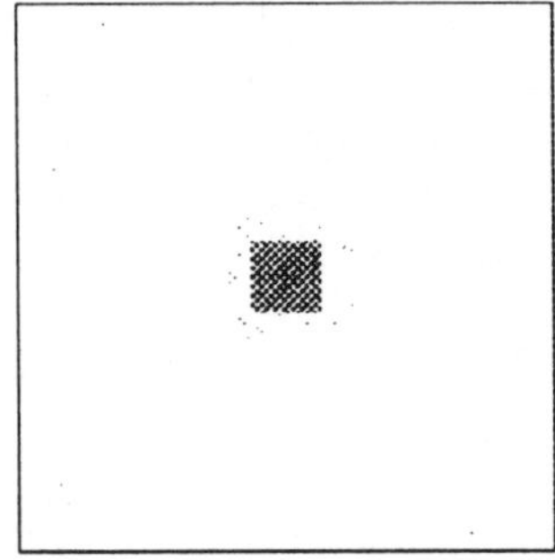

Figure 7.12 The intensity distribution in a lens focus obtained from the DOE with a phase shown in Fig. 7.11

Table 7.4 Comparison of DOEs

Type of DOE and phase	Energy efficiency E (%)	Relative r.m.s. error δ (%)
Ray optics phase	91.6	5.0
10 iterations of the Gerchberg–Saxton algorithm on the ray optics phase	98.9	13.0
10 iterations of the adaptive–iterative algorithm on the ray optics phase	92.2	0.1
10 combined iterations on the ray optics phase	96.4	2.4
10 iterations on the radius-random phase	80.8	6.4

ray optics phase (compare the first four rows with the fifth). Comparison between the first and second rows in the table shows that the Gerchberg–Saxton method yields an increase of 7% in the energy efficiency of the ray optics approximation (first row), but produces considerable nonuniformity of the intensity over the square (13% instead of 5%).

On the contrary, the adaptive method ensures a high degree of intensity uniformity (0.1%) but practically does not improve the efficiency as compared with the ray optics approximation (third row). The combined method enables us to obtain rather high efficiency in combination with satisfactory intensity uniformity (fourth row in the table). The fifth row shows that not knowing the ray optics solution but based upon the random initial phase in the estimate of the DOE's phase and using the iterative method, we can achieve suitable results in terms of accuracy and efficiency in the formation of the required intensity distribution.

Thus, we have shown numerically that an optical approach to a problem of computation of phase optical elements focusing the laser light into the small areas of the spatial spectrum plane involves solving the inverse ray optics task and finding the phase function which is then chosen as the initial approximation for the adaptive–iterative procedure of obtaining the final phase. In this case, the iterative procedure of the correction for the initial phase with regular zone structure does not result in its essential change.

7.3 Focusing into a Ring

With the aim of focusing the coherent light at a narrow ring, cone axicons matched to spherical lenses [87] and binary axicons [151] are commonly used. If a term linearly depending on the azimuth angle is added to the phase function describing the transmittance of the axicon, the resultant optical element will possess new properties [152]. In section 7.3.1 we discuss the operation of the axicon combined with a rotor component. For example, in the course of forming a ring it is usual that the light spot occurs on the optical axis, even though the element has been synthesized without errors. As will be apparent below [50], the rotor component, when added to the axicon, partially solves the problem.

In section 7.3.2, with the aim of focusing into a wide ring with a preset intensity distribution along the annulus radius, we consider a method for calculating a DOE aimed at focusing into a circular off-axis domain. The method combines the reduction of the problem of focusing into a radial off-axis domain to a 1D problem of focusing into a segment and the use of iterative techniques for solving a 1D task of focusing into a segment. The method proposed needs only two Fourier transforms to be taken for each iteration, thus reducing the computational effort by a factor of three as compared with the iterative calculation via the Hankel transform described in Chapter 2.

7.3.1 Focusing into a Narrow Ring

By a narrow light ring in the focal plane [84], we mean one whose width is essentially the Fraunhofer diffraction limit of the finite-aperture conical wave diffracted by the DOE.

To focus into a narrow ring, one can employ an optical element – a cone axicon with the complex transmission function given by

$$\tau(r) = \exp(-i\alpha r) \tag{7.24}$$

where α is the characteristic of the axicon 'power' and r is the radial coordinate. If the element given by Eq. (7.24) is illuminated by the plane monochromatic light beam with the wavenumber $k = 2\pi/\lambda$ (where λ is the wavelength), and a spherical lens of focal length f is placed behind the element, in this case in the rear focal plane of that lens the light ring of radius $R = \alpha f/k$ and width $b = \lambda f/a$ will be formed, where a is the radius of the axicon aperture. The light complex amplitude in the lens focal plane is given by the Fraunhofer diffraction integral expressed in polar coordinates

$$F_0(\rho) = \frac{k}{f}\int_0^a \mathrm{e}^{-i\alpha r} J_0\left(\frac{k}{f} r\rho\right) r\,\mathrm{d}r \tag{7.25}$$

where $J_0(x)$ is the Bessel function of zero order. The light amplitude on the ring is computed by the formula

$$\begin{aligned} F_0(R) &= \frac{k}{f}\int_0^a \mathrm{e}^{-i\alpha r} J_0(\alpha r)\, r\,\mathrm{d}r \\ &= \frac{k}{f\alpha^2}\mathrm{e}^{i\xi}\left\{\left[\xi J_1(\xi) - \frac{\xi^2}{3}J_2(\xi)\right] + i\frac{\xi^2}{3}J_1(\xi)\right\} \end{aligned} \tag{7.26}$$

where $\xi = \alpha a$, and $J_1(x)$ and $J_2(x)$ are the Bessel functions of the first and second orders, respectively. The light intensity in the centre of the ring is computed by

$$I_0(0) = \left|\frac{k}{f}\int_0^a \mathrm{e}^{-i\alpha r} r\,\mathrm{d}r\right|^2 = \left(\frac{ka^2}{2f}\right)^2\left[\left(\frac{\sin\nu}{\nu}\right)^2 + \left(\frac{\cos\nu}{\nu} - \frac{\sin\nu}{\nu^2}\right)^2\right] \tag{7.27}$$

where $\nu = \alpha a/2$.

As one can see from the last term in Eq. (7.27), the intensity in the centre of the ring assumes its maximum value for $\alpha = 0$

$$I_0(0)|_{\max} = \left(\frac{ka^2}{2f}\right)^2 \tag{7.28}$$

The intensity in the ring centre asymptotically attains its zero value when α tends to infinity. We can find the values that give us the local maxima and minima of the intensity as the solutions of the equation

$$\frac{\mathrm{d}I_0}{\mathrm{d}\nu} = \left(\frac{ka^2}{2f}\right)^2\left[\frac{-4}{\nu}\left(\frac{\cos\nu}{\nu} - \frac{\sin\nu}{\nu^2}\right)^2\right] = 0 \tag{7.29}$$

whose solution is

$$\nu = \tan \nu \tag{7.30}$$

The energy efficiency of focusing into a narrow ring based on the use of the axicon is found from

$$\varepsilon_0 = 2\pi R b I_0(R)[2\pi R b I_0(R) + \pi b^2 I_0(0)]^{-1} \tag{7.31}$$

where $b = \lambda f/a$ is the ring width which is almost equal to the radius of the central light spot, and $I_0(R) = |F_0(R)|^2$,$F_0(R)$ is taken from Eq. (7.26) and given by

$$I_0(R) = \left(\frac{k}{f}\right)^2 \frac{a^4}{9}[J_1^2(\xi) + J_2^2(\xi)] + \left(\frac{k}{f}\right)^2 a^4 \frac{J_1(\xi)}{\xi}\left(\frac{J_1(\xi)}{\xi} - \frac{2}{3}J_2(\xi)\right) \tag{7.32}$$

To focus into the narrow ring, we can also use the rotor axicon with the transmission function as [50,84]

$$\tau(r, \varphi) = \exp[-i\alpha r + im\varphi] \tag{7.33}$$

where $m = 0, 1, 2, \ldots$, and φ is the azimuth angle. Note that α can take both positive and negative values. The complex light amplitude in the lens focal plane is described in this case instead of Eq. (7.17) by

$$F_m(\rho, \psi) = \frac{k}{f} i^m e^{im\psi} \int_0^a e^{-i\alpha r} J_m\left(\frac{k}{f} r\rho\right) r\,dr \tag{7.34}$$

where $J_m(x)$ is the Bessel function of the mth order and (ρ, ψ) are the polar coordinates in the Fourier plane. It is seen from Eq. (7.34) that regardless of the value of α and a, in the centre of the ring the intensity equals zero, $J_m(0) = 0$, for $m > 0$. With an increase of the order m, the intensity in the vicinity of the ring centre falls. The intensity on the ring is computed by $(m = 1)$

$$I_1(R) = |F_1(R, \psi)|^2 = \left(\frac{k}{f}\right)^2 \frac{a^4}{9}[J_1^2(\xi) + J_2^2(\xi)] \tag{7.35}$$

One can see from comparison of Eqs (7.32) and (7.35) that, when one is using the rotor axicon of Eq. (7.33), the intensity of the light striking the ring can be either greater or smaller than that upon employing the axicon of Eq. (7.24) and depends on the sign of the second term in the expression

$$I_0(R) = I_1(R) + \left(\frac{k}{f}\right)^2 a^4 \frac{J_1(\xi)}{\xi}\left(\frac{J_1(\xi)}{\xi} - \frac{2}{3}J_2(\xi)\right)$$

But the rotor axicon does not give significant benefit in the energy efficiency as compared with the axicon.

Figure 7.13a,c illustrates the phase modulo 2π for the axicon and the rotor axicon, respectively. Figure 7.14a,c shows the intensity distributions produced by these optical elements in the lens focal plane. It is seen that

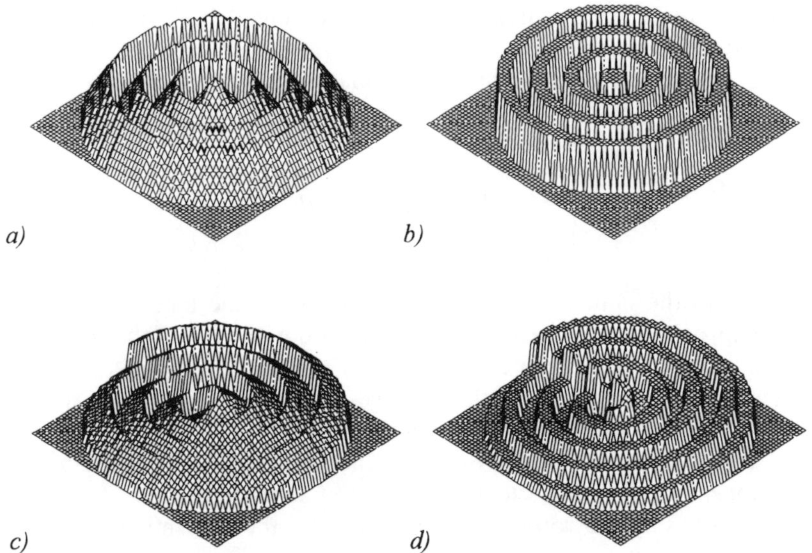

Figure 7.13 Phases of the DOE into a narrow ring: (a) axicon; (b) binary axicon; (c) rotor axicon; (d) rotor binary axicon

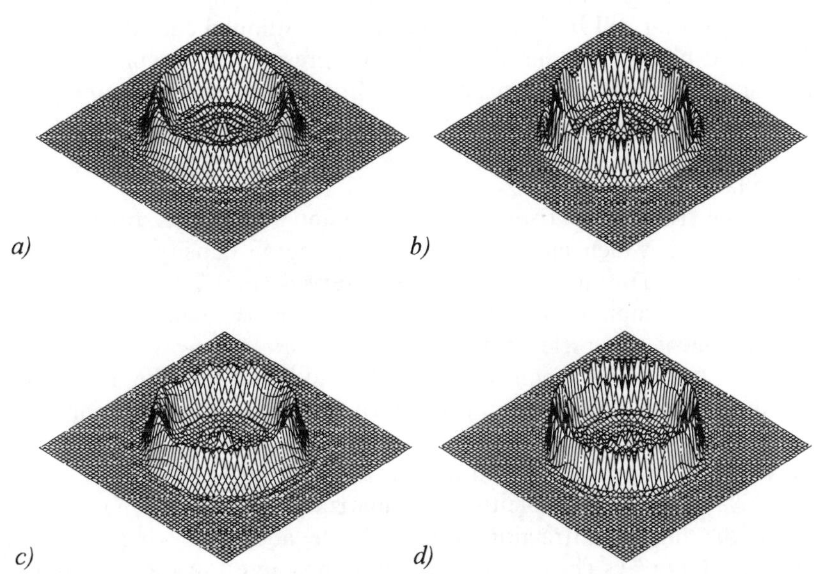

Figure 7.14 Intensity distributions in the lens focal plane: (a) for the axicon; (b) for the binary axicon; (c) for the rotor axicon; (d) for the rotor binary axicon

if we use the rotor axicon, instead of the 'peak' (Fig. 7.14a) there is a 'cone' in the centre of the ring (Fig. 7.14c).

Let us consider the result of the axicon phase binarization that is employed to put the optical element into a convenient form for manufacture. The binary axicon with a transmission function of

$$\tau(r) = \begin{cases} 1, & 2\pi k \leq |\alpha r| \leq \pi(2k+1) \\ -1, & \pi(2k+1) < |\alpha r| < 2\pi(k+1) \end{cases} \tag{7.36}$$

will also form the light ring in the Fourier plane of the lens. And the ring width is approximately twice as small than that upon the use of the continuous axicon of Eq. (7.24). This is due to the fact that in each point of the light ring there is an interference of the positive and negative orders of the light transmitted by the binary radial grating of Eq. (7.36). Also, such interference results in the increase of the undesirable modulation in the intensity distribution along the ring when changing the azimuth angle.

If we synthesize the rotor binary axicon with the transmission function as

$$\tau(r, \varphi) = \begin{cases} \exp(im\varphi), & 2\pi k \leq |\alpha r| \leq \pi(2k+1) \\ -\exp(im\varphi), & \pi(2k+1) < |\alpha r| < 2\pi(k+1) \end{cases} \tag{7.37}$$

and in each point of the ring the first positive and negative orders of the light diffracted by the grating of Eq. (7.37) interfere in the negative, a split (double) light ring will be formed. The above-mentioned considerations can be illustrated by an example. Figure 7.13b,d presents the arguments of the transmission function of the binary axicon, Eq. (7.36), and the rotor (helical) binary axicon, Eq. (7.37), respectively. Figure 7.14b,d illustrates corresponding light intensity distributions produced by these optical elements in the Fourier plane of the lens.

As is seen from comparison of Figs 7.14a and 7.14b, a narrowing of the ring takes place, which means that the light energy density on the preset radius increases. The angular variations observed (Fig. 7.14b) are a result of inadequate sampling: a small number of pixels leads to the rough approximation of the circle by a polygon.

It is seen from comparison of Figs 7.14a and 7.14d that the binarization of the rotor axicon causes differentiation of the intensity on the ring, i.e. a double ring appears.

The energy efficiency of focusing in Fig. 7.14a–d equals, respectively, 87%, 83.6%, 90%, and 83.3%. Note that the binarization of the axicon brings into existence the higher diffraction orders. These additional rings were not included in the arrays chosen and were not considered in the computation of the energy efficiency. Therefore, the foregoing values of efficiency turn out to be too high and are meaningful only in terms of their comparison with one another.

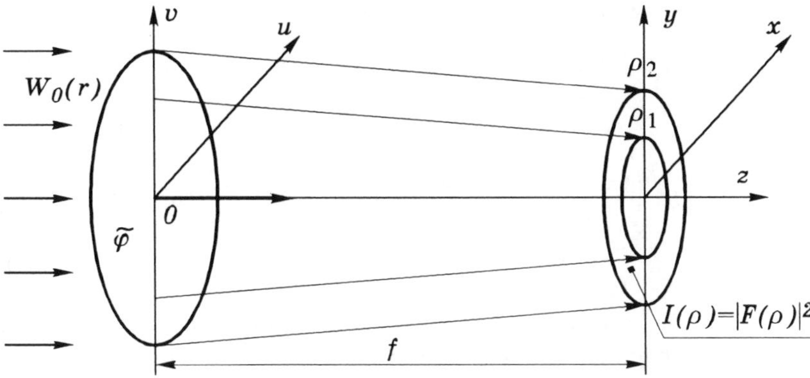

Figure 7.15 Geometry of the problem of focusing into a circular domain

7.3.2 *Focusing into a Circular Domain*

7.3.2.1 *Reducing the problem of focusing into a circular off-axis domain to focusing into a segment*

Let us assume that a laser beam with the complex amplitude given by

$$W_0(r) = \sqrt{[I_0(r)]}\exp[i\varphi_0(r)]$$

where $I_0(r)$ is the illuminating beam intensity and $\varphi_0(r)$ is the beam phase, strikes a DOE characterized by the complex transmission $\exp[i\tilde{\varphi}(r)], r \leq a$. It is desired to find the DOE phase function $\tilde{\varphi}(r)$ providing the generation of a pregiven circular intensity distribution $I(\rho), \rho_1 \leq \rho \leq \rho_2$, in the focal plane $z = f$ (see Fig. 7.15).

In what follows, the DOE phase function will be considered to take the form

$$\tilde{\varphi}(r) = \varphi(r) - \varphi_0(r) \tag{7.38}$$

The representation of Eq. (7.38) allows us to calculate the DOE without regard for the illuminating beam phase. The field complex amplitude in the DOE focal plane is given by the Fresnel–Kirchhoff integral expressed in polar coordinates

$$F(\rho) = \frac{k}{f}\exp\left(i\frac{k\rho^2}{2f}\right)\int_0^a \sqrt{[I_0(r)]}\exp[i\varphi(r)]\exp\left(i\frac{kr^2}{2f}\right)J_0\left(\frac{k}{f}r\rho\right)r\,\mathrm{d}r \tag{7.39}$$

Using the asymptotic representation of $J_0(\xi)$ at $\xi \gg 0$ [110]

$$J_0(\xi) = \sqrt{\left(\frac{2}{\pi\xi}\right)}\cos\left(\xi - \frac{\pi}{4}\right), \quad \xi \to \infty \tag{7.40}$$

gives the following approximation for $F(\rho)$ at $\rho \gg 0$

$$F(\rho) = \exp(-i\pi/4)[F_1(\rho)/\sqrt{(\rho)} + iF_2(\rho)/\sqrt{(\rho)}] \quad (7.41)$$

where

$$F_1(\rho) = \sqrt{\left(\frac{k}{2\pi if}\right)} \int_0^a \sqrt{[I_0(r)r]} \exp[i\varphi(r)] \exp\left(\frac{ik(\rho - r)^2}{2f}\right) \mathrm{d}r \quad (7.42)$$

$$F_2(\rho) = F_1(-\rho) \quad (7.43)$$

The $F_1(\rho)$ function corresponds to the field complex amplitude formed by a 1D DOE illuminated by a beam with the intensity distribution

$$\tilde{I}(r) = I_0(r)r \quad (7.44)$$

The problem of focusing into a radial domain with intensity distribution $I(\rho)$ for $\rho \gg 0$ can be reduced to a 1D problem of focusing into a segment. Let $\varphi(r)$ be the phase function of a 1D DOE focusing a beam of Eq. (7.44) into a segment with the intensity distribution

$$I_1(\rho) = I(\rho)\rho, \quad \rho_1 \leq \rho \leq \rho_2 \quad (7.45)$$

According to Eq. (7.43), the term $F_2(\rho)$ corresponds to the field complex amplitude upon focusing into a segment at $-\rho_2 \leq \rho \leq -\rho_1$, and does not affect the field $F_1(\rho)$ for $\rho_1 \leq \rho \leq \rho_2$ (see Fig. 7.16).

Therefore, the approximation for $F(\rho)$ takes the form

$$F(\rho) \approx \exp(-i\pi/4) F_1(\rho)/\sqrt{(\rho)} \quad (7.46)$$

Accordingly, the function $|F_1(\rho)|^2/\rho$ representing the intensity distribution in the focal plane of the radial DOE turns into a required intensity distribution $I(\rho)$ for $I_1(\rho) = |F_1(\rho)|^2$ given in Eq. (7.45).

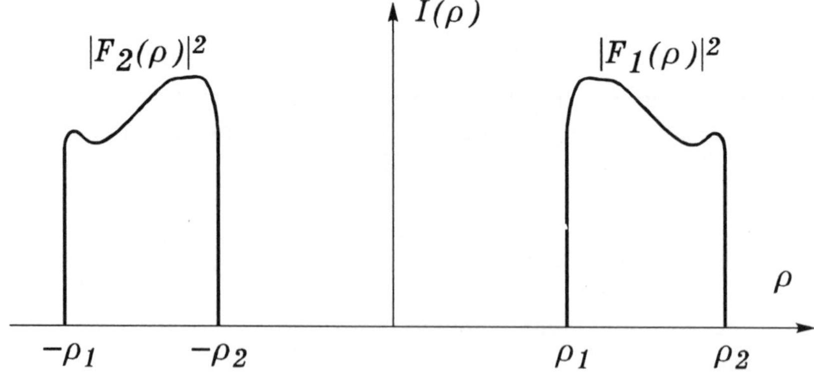

Figure 7.16 The field structure described by the terms $F_1(\rho)$ and $F_2(\rho)$

It is noteworthy that in calculating the integral (7.42) using a stationary phase method, the approximation of Eq. (7.46) leads to the familiar formulae for the focusator phase function [37,84,126]

$$\varphi(r) = -\frac{kr^2}{2f} + \frac{k}{f}\int_0^r \tilde{\rho}(\xi)\,\mathrm{d}\xi, \quad r \in [0, a] \tag{7.47}$$

where the $\tilde{\rho}(\xi)$ function is derived from

$$\int_0^r I_0(\zeta)\,\zeta\,\mathrm{d}\zeta = \int_{\rho_1}^{\tilde{\rho}(r)} I(\xi)\,\xi\,\mathrm{d}\xi \tag{7.48}$$

In Refs [37,84], Eq. (7.47) is derived from the eikonal equation, and Eq. (7.48) is deduced from the transfer equation and corresponds to the integral form of the light flux conservation law.

For example, the phase function of a DOE aimed at focusing a plane beam into a wide ring with a uniform intensity distribution can be derived from Eqs (7.47) and (7.48) in the form

$$\varphi(r) = -\frac{kr^2}{2f} + \frac{k\rho_1}{2fc}\{cr(c^2r^2+1)^{1/2} + \ln[cr + (c^2r^2+1)^{1/2}]\} \tag{7.49}$$

where $c = (\rho_2^2 - \rho_1^2)^{1/2}/(a\rho_1)$, and ρ_1 and ρ_2 are the inside and outside radii of the focal ring, respectively.

7.3.2.2 *Estimation of the approximation accuracy*

Consider the accuracy of the obtained approximation of Eqs (7.42) and (7.46) using a model example of focusing into a narrow ring. For $\rho_1 \to \rho_2$, the function $\varphi(r)$ of Eq. (7.49) corresponds to the superposition of the phase functions of a thin lens and an axicon ($c \to 0$)

$$\varphi(r) = -\frac{kr^2}{2f} + \frac{k}{f}\rho_1 r \tag{7.50}$$

and provides focusing into a narrow ring of radius ρ_1.

Let $I(\rho_1)$ and $\hat{I}(\rho_1)$ be the intensities on the geometrical ring that have been deduced, respectively, through the Fresnel–Kirchhoff integral, Eq. (7.39), and using the approximation of Eqs (7.42) and (7.46). To find the limits within which the approximation of Eqs (7.42) and (7.46) is feasible, let us compare $I(\rho_1)$ and $\hat{I}(\rho_1)$ for different values of ρ_1, i.e. for different distances from the focal ring to the optical axis. As a criterion, let us use the value of the relative error

$$\varepsilon(\rho_1) = \frac{|\hat{I}(\rho_1) - I(\rho_1)|}{I(\rho_1)}$$

Table 7.5 Approximation error $\varepsilon(\rho_1)$ versus $h = \rho_1/\Delta$

h	$\varepsilon(\rho_1)$ (%)
0.5	48.1
1.0	26.5
1.5	17.2
2.0	13.1
2.5	10.4
3.0	7.9
3.5	5.7
4.0	4.9
4.5	4.8
5.0	3.5

For $\varphi(r)$ in Eq. (7.50), we obtain from Eqs (7.42) and (7.46) the intensity on the geometric ring (for $\rho = \rho_1$) in the form

$$\hat{I}(\rho_1) = \frac{2ka^3}{9\pi f \rho_1} \tag{7.51}$$

The intensity $I(\rho_1)$ value derived from the Fresnel–Kirchhoff integral is given in Eq. (7.32). Table 7.5 summarizes the calculated values of $\varepsilon(\rho_1)$ for a focusator focusing into a narrow ring for different values of the parameter $h = \rho_1/\Delta$, $\Delta = \lambda f/a$. The quantity h characterizes the size of the focal ring radius in comparison with the diffraction size Δ. $I(\rho_1)$ and $\hat{I}(\rho_1)$ were calculated via Eqs (7.51) and (7.32) for the following parameters: $\lambda = 1.06\,\mu$m, $a = 2.5$ mm, and $f = 250$ mm.

As is seen from Table 7.5, the error $\varepsilon(\rho_1)$ for $h > 4$ is less than 5%, indicating good accuracy of the approximation developed.

7.3.2.3 *Calculation of DOEs focusing into a wide ring*

As demonstrated in section 7.3.2.1, the problem of calculating a DOE for focusing into a radial off-axis circular domain reduces to calculating the phase function $\varphi(r)$ of a 1D DOE focusing into a segment. The calculational results of section 7.3.2.2 allow definition of the focal off-axis domain as a domain whose inside radius is 4–5 times greater than the diffraction size $\Delta = \lambda f/a$.

A ray-tracing method for calculating $\varphi(r)$ leads to the familiar Eqs (7.47) and (7.48) for the focusator phase function. The calculation of $\varphi(r)$ via a 1D iterative algorithm corresponds to the new iterative method for calculating a DOE focusing into a circular off-axis domain.

Chapter 2 discusses the iterative calculation of DOEs aimed at focusing into a radial domain. In Chapter 2 the recalculation of the field between the DOE location and focal planes is implemented via the Hankel transform.

The Hankel transform realization is reduced to taking three Fourier transforms with the help of the exponential replacement of variables. Therefore, the iterative technique described in Chapter 2 calls for six Fourier transforms in each iteration step. The iterative method developed in this section needs two Fourier transforms to be taken per iteration, i.e. three times less than with the DOE iterative calculation via Hankel transforms. In the following, the Hankel transform-based iterative calculation will be called a radial iterative calculation. The calculation of a radial DOE based on the iterative calculation of a DOE for focusing into a segment will be called a linear iterative calculation. In the present section, we compare the DOEs focusing into a wide ring that have been designed using a ray-tracing approach for the derivation of $\varphi(r)$ (see Eq. (7.49)), a linear iterative calculation of $\varphi(r)$, and a radial iterative calculation of $\varphi(r)$.

To estimate the DOE performance, let us use the energy efficiency E and the r.m.s. deviation δ. The quantity

$$E = \int_{\rho_1}^{\rho_2} I(\rho)\rho \mathrm{d}\rho \bigg/ \int_0^a I_0(r) r \mathrm{d}r$$

characterizes the portion of the illuminating beam energy which is focused into a given focal domain. The quantity

$$\delta = \frac{1}{\bar{I}} \sqrt{\left(\frac{2}{(\rho_2^2 - \rho_1^2)} \int_{\rho_1}^{\rho_2} [I(\rho) - \bar{I}]^2 \rho \mathrm{d}\rho \right)}$$

characterizes the r.m.s. deviation of the intensity distribution from an average value

$$\bar{I} = \frac{2}{(\rho_2^2 - \rho_1^2)} \int_{\rho_1}^{\rho_2} I(\rho)\rho \mathrm{d}\rho$$

The linear and radial iterative calculations of $\varphi(r)$ were conducted using the AA algorithm described in Chapter 1. The phase function of Eq. (7.49) obtained by the ray-tracing method was chosen as the starting-point for the iterative process. The recalculation of the field between the planes was conducted via a fast Fourier transform with the number of pixels $N = 256$ and for the following parameters: $\lambda = 1.06\,\mu\mathrm{m}$, $a = 2.5$ mm, and $f = 250$ mm. Table 7.6 gives the values for E and δ versus the parameter $S = (\rho_2 - \rho_1)/\Delta, \Delta = \lambda f/a$, for $\rho_1 = 28\Delta$. The quantity S characterizes the width of the focal ring in comparison with the diffraction size Δ.

From the data given in Table 7.6, it is seen that the r.m.s. error δ for the DOEs calculated using the linear iterative algorithm is 3–5 times less than that for focusators, with the energy efficiency E differing insignificantly. The radial iterative algorithm does not provide an essential improvement of E and δ as compared with the DOEs calculated via the linear algorithm.

Figure 7.17 shows phase functions of the focusator and of the DOEs calculated using the linear and radial algorithms, respectively, for $S = 10$. The

Table 7.6 The E and δ parameters for the focusator and DOE calculated using the linear and radial iterative algorithms

	Ray-tracing method		Linear iterative algorithm		Radial iterative algorithm	
S	E (%)	δ (%)	E (%)	δ (%)	E (%)	δ (%)
8	90.4	48.0	89.1	13.1	89.2	13.3
10	89.9	39.4	92.5	11.1	92.6	11.0
12	90.7	37.4	89.4	9.7	89.5	9.6
14	90.9	40.0	91.3	9.1	91.5	9.2
16	90.2	37.7	90.4	7.8	90.7	7.7
18	89.7	34.5	90.8	7.4	90.7	7.2
20	89.7	32.4	90.1	6.9	90.2	6.6

calculated intensity distributions in Fig. 7.18 reveal an almost total absence of the intensity distribution fluctuations within the focal ring for the 'linear' and 'radial' DOEs. The focusator and linear DOE create intensity peaks at $\rho = 0$. The peak's presence is explained by the fact that both the ray-tracing method and the linear iterative algorithm do not control the intensity distribution near the optical axis. For the radial DOE, the central peak did not appear. The absence of a central peak explains the significant difference between the linear and radial DOEs phase functions shown in Figs 7.17b and 7.17c.

7.4 Focusing into an Axial Segment

In such areas as optical gas discharge [69], optical disk systems [88], and devices for non-contact displacement analysis [89], there is a growing demand for optical elements focusing laser radiation into an axial segment. In this section we consider analytic and iterative methods for calculating DOEs capable of focusing the radiation into an optical axis segment.

Assume that a laser beam with the complex amplitude given by

$$W_0(r) = \sqrt{[I_0(r)]} \exp[i\varphi_0(r)], \quad r \in [0, R] \tag{7.52}$$

where $I_0(r)$ is the illuminating beam intensity and $\varphi_0(r)$ is the beam phase, strikes a DOE with a round aperture of radius R. The DOE locates at the plane $z = 0$ (Fig. 7.19) and transforms the incident radiation into a field characterized by the complex amplitude

$$W(r) = W_0(r) \exp[i\hat{\varphi}(r)] \tag{7.53}$$

where $\hat{\varphi}(r)$ is the DOE phase function.

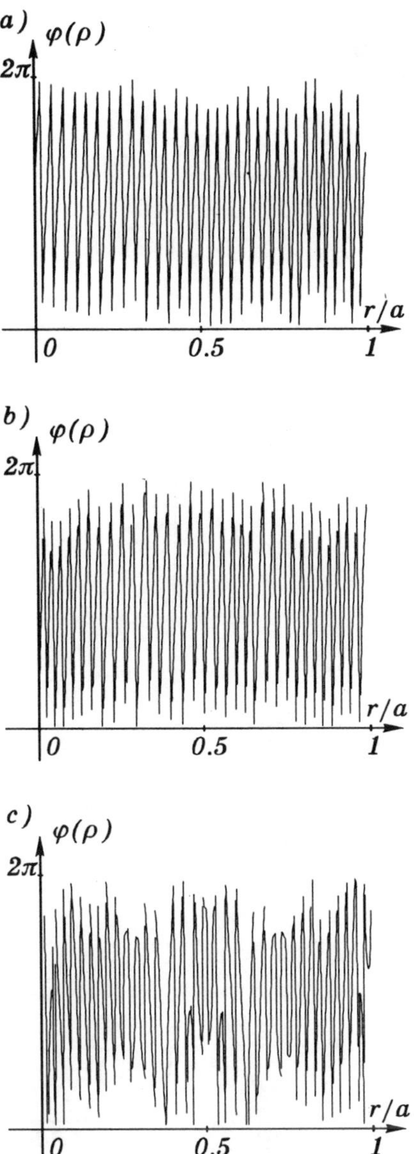

Figure 7.17 Phase functions (a) for the focusator and (b) for the DOE calculated using the linear iterative algorithm, and (c) for the DOE calculated using the radial iterative algorithm

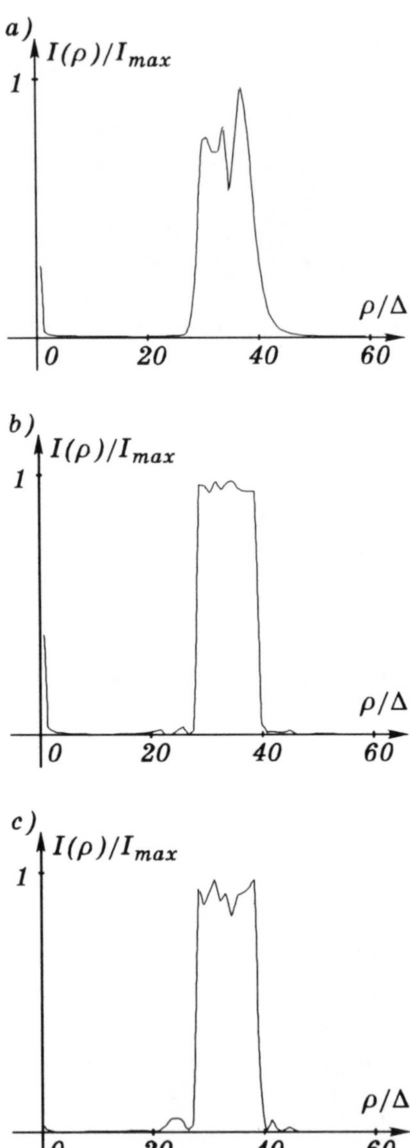

Figure 7.18 Intensity distributions in the focal plane (a) for the focusator, (b) for the DOE calculated using the linear iterative algorithm, and (c) for the DOE calculated using the radial iterative algorithm

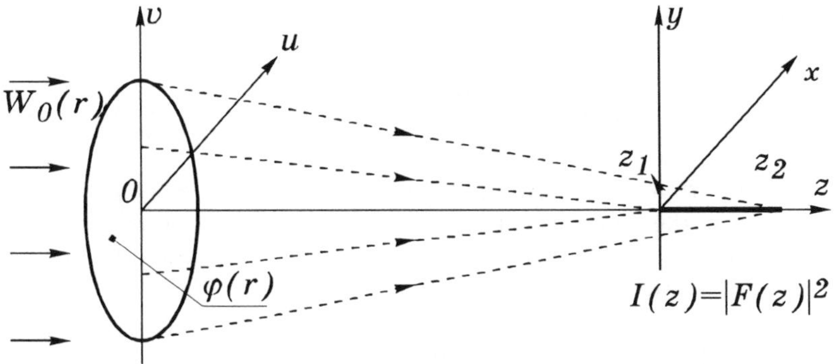

Figure 7.19 Geometry of the problem of focusing into an axial segment

It is necessary to calculate the DOE phase function $\hat{\varphi}(r)$ providing the focusing of the incident beam into an optical axis segment with the intensity distribution $I(z), z \in [z_1, z_2]$.

For convenience, we represent the $\hat{\varphi}(r)$ function in the form

$$\hat{\varphi}(r) = \varphi(r) - \varphi_0(r) \tag{7.54}$$

The representation of Eq. (7.54) makes it possible to calculate the DOE irrespective of the illuminating beam phase $\varphi_0(r)$. In the subsequent discussion, we shall operate with the $\varphi(r)$ function that uniquely determines the DOE's phase transmittance.

The field complex amplitude $F(\rho, z)$, where $\rho = \sqrt{(x^2 + y^2)}$, is specified by the DOE complex transmittance $\exp[i\varphi(r)]$ through the Fresnel–Kirchhoff integral expressed in polar coordinates

$$F(\rho, z) = \frac{k}{z}\exp\left(i\frac{k\rho^2}{2z}\right)\int_0^R \sqrt{[I_0(r)]}\exp[i\varphi(r)]\exp\left(i\frac{kr^2}{2z}\right) \times J_0\left(k\frac{r\rho}{z}\right) r\,\mathrm{d}r \tag{7.55}$$

where $k = 2\pi/\lambda$, λ is the wavelength and $J_0(\rho)$ is the Bessel function of zero order.

For the field complex amplitude on the optical axis ($\rho = 0$), Eq. (7.55) takes the form

$$F(z) = \frac{k}{z}\int_0^R \sqrt{[I_0(r)]}\exp\left(i\varphi(r) + ik\frac{r^2}{2z}\right) r\,\mathrm{d}r \tag{7.56}$$

Substituting the variables in the form

$$\xi = -\frac{z_1}{z} \quad \text{and} \quad x = \frac{r^2}{2} \tag{7.57}$$

allows Eq. (7.56) to be rewritten as

$$F(-z_1/\xi) = -\frac{k}{z_1}\,\xi \int_0^{R^2/2} \surd[I_0 \surd(2x)] \exp[i\varphi_1(x)] \exp\left(-ik\frac{x\xi}{z_1}\right) dx \tag{7.58}$$

where $\varphi_1(x) = \varphi\surd(2x)$.

It follows from Eq. (7.58) that the problem of focusing into an axial segment with the intensity distribution $I(z)$, $z \in [z_1, z_2]$ reduces to the calculation of the phase function $\varphi_1(x)$ of a 1D DOE focusing the beam with the complex amplitude

$$\hat{W}_0(x) = \surd[I_0 \surd(2x)] \exp\left(-ik\frac{x^2}{2z_1}\right), \quad x \in \left[0, \frac{R^2}{2}\right]$$

into a transverse segment with the intensity distribution $I_1(\xi) = I(-z_1/\xi)/\xi^2, \xi \in [-1, -z_1/z_2]$ in a plane placed at distance z_1 from the 1D DOE aperture. Using a variety of techniques for calculating the 1D DOE phase function $\varphi_1(x)$, we can derive a variety of solutions of the problem of focusing into an axial segment.

7.4.1 Calculation of Focusators and Quasi-periodic DOEs

Based on the ray-tracing approach, the phase function $\varphi_1(x)$ of a 1D DOE is sought from the solution of the following set of equations [37,153]

$$\begin{cases} \varphi_1(x) = \varphi(x) - \varphi_0(x) \\ \dfrac{d\varphi(x)}{dx} = \dfrac{k}{z_1}[\xi(x) - x] \\ \dfrac{\hat{I}_0(x)}{I_1(\xi)} = \dfrac{d\xi(x)}{dx}, \quad x \in [x_0, x_1]; \xi \in [\xi_0, \xi_1] \end{cases} \tag{7.59}$$

where $\varphi_0(x)$ and $\hat{I}_0(x)$ are the phase and intensity of the illuminating beam.

In particular, when focusing the plane beam into an axial segment with a constant intensity, Eqs (7.57) and (7.59) give the phase function of a DOE focusing into an axial segment in the form

$$\varphi(r) = \frac{-kR^2}{2L} \ln\left[\frac{r^2 L}{R^2 z_1} + 1\right] \tag{7.60}$$

where $L = z_2 - z_1$.

For $L/z_2 \ll 1$, the intensity distribution $I_1(\xi)$ is close to uniform. In this case, Eqs (7.57) and (7.59) reduce the DOE phase function to the form

$$\varphi(r) = \frac{-kr^2}{2z_1} + \frac{kr^4 L}{4R^2 z_1 z_2} \tag{7.61}$$

The phase function of Eq. (7.61) corresponds to the phase function of a thin lens with introduced spherical aberration.

One should note that the ray-tracing approach to calculating $\varphi_1(x)$ gives Eqs (7.60) and (7.61) that are equivalent to the phase functions of focusators focusing into an axial segment reported in [33,152,154]. In this connection, in what follows we shall call the DOEs of Eqs (7.60) and (7.61) focusators.

It is also possible to find the phase function $\varphi_1(x)$ through the use of the iterative algorithms, allowing the improvement of the ray-tracing solution of Eqs (7.60) and (7.61).

Reference [79] reports the calculation of 'quasi-periodic' DOEs similar to holograms in their functional properties, in that every part of the DOE's aperture forms the entire focal domain. In the case of a converging illuminating beam, the phase function $\varphi_1(x)$, $x \in [x_0, x_1]$ of a 1D quasi-periodic DOE corresponds to a K-times repeated phase function $\varphi_p(x)$, $x \in [0, (x_1 - x_0)/K]$ which ensures focusing into desired line segments [79]. In each repetition, a constant phase shift

$$\varphi_i = \frac{\pi i^2}{K}, \quad i = 0, \ldots, K-1, K \text{ even} \tag{7.62}$$

is introduced [79]. Thus, the phase function of a quasi-periodic DOE takes the form:

$$\varphi_1(x) = \varphi_p[\operatorname{mod}_\alpha(x - x_0)] + \frac{\pi}{K}\{\operatorname{int}[(x - x_0)/\alpha]\}^2 \tag{7.63}$$

where $\alpha = (x_1 - x_0)/K$ and $\operatorname{int}[x]$ is the integral part of x.

The function $\varphi_p(x), x \in [0, \alpha]$ in Eq. (7.63) can be derived from Eq. (7.59) or calculated using an iterative Gerchberg–Saxton algorithm [14].

Using the $\varphi_p(x)$ function derived from Eq. (7.59), one can find the phase function of a quasi-periodic DOE focusing the plane beam into an axial segment in the form

$$\varphi(r) = \frac{-kR^2}{2KL} \ln\left(\frac{2LK}{R^2 z_1} \operatorname{mod}_\alpha(r^2/2) + 1\right) + \tilde{\varphi}(r) \tag{7.64}$$

where $\alpha = R^2/2K$

$$\tilde{\varphi}(r) = \frac{\pi}{K}(\operatorname{int}[r^2/2\alpha])^2 \tag{7.65}$$

If $L/z_2 \ll 1$, the DOE phase function can be reduced to the following form

$$\varphi(r) = -\frac{k}{z_1} \operatorname{mod}_\alpha[r^2/2] + \frac{kLK}{R^2 z_1 z_2}[\operatorname{mod}_\alpha(r^2/2)]^2 + \tilde{\varphi}(r) \tag{7.66}$$

Note that for $K = 1$, Eqs (7.64) and (7.66) are equivalent to the phase function of the focusators of Eqs (7.60) and (7.61). For the quasi-periodic

DOE specified by Eq. (7.64) or (7.66), the $\varphi_p(x)$ function in Eq. (7.63) corresponds to focusing into a transverse line segment of length $N(K)\Delta$, where $\Delta = 2\lambda z_1 K/R^2$ is the size of diffractive resolution, and

$$N(K) = \text{int}\left[\frac{LR^2}{2\lambda z_1 z_2 K}\right] \tag{7.67}$$

The use of the ray-tracing relations of Eq. (7.59) for the calculation of the function $\varphi_p(x)$ in Eq. (7.63) is feasible only for $N(K) \gg 1$ (e.g. for $N \geq 10$) imposing restrictions on possible values of K. If $N(K) = 1 \div 3$, the calculation of a quasi-periodic DOE appears to be impossible even with iterative calculation of the $\varphi_p(x)$ function in Eq. (7.63).

Let us analyze Eq. (7.66). If $\tilde{\varphi}(r) \equiv 0$, the phase function in Eq. (7.66) goes over to the phase function of a multifocus lens with $N(K)$ foci. Actually, using the relation

$$\text{mod}_\alpha[f(x)] = \frac{\alpha}{2\pi}\,\text{mod}_{2\pi}[2\pi f(x)/\alpha]$$

Eq. (7.66) can be recast in the form

$$\varphi(r) = \Phi[\varphi_2(r)] \tag{7.68}$$

where

$$\Phi[x] = \frac{kLR^2}{16K\pi^2 z_1 z_2}x^2 - \frac{kR^2}{4K\pi z_1}x, \quad x \in [0, 2\pi] \tag{7.69}$$

$$\varphi_2(r) = \text{mod}_{2\pi}\left(-\frac{kr^2}{2f_2}\right), \quad f_2 = -\frac{kR^2}{4\pi K} \tag{7.70}$$

Equations (7.68)–(7.70) correspond to the phase function of the multifocus lens given by Eqs (6.58) and (6.59), at $f_1 = \infty$. From Eq. (7.59), one can easily find that the function $\Phi[x]$ specified by Eq. (7.69) is the phase function of a focusator aimed at focusing the converging cylindrical beam into a transverse line segment

$$-\frac{R^2}{4\pi K} \leq \xi \leq -\frac{R^2 z_1}{4\pi K z_2}$$

with uniform intensity distribution at the plane $z = f_1$. In Chapter 5, it is shown that the function $\Phi[x]$ corresponds to the phase function of a diffraction grating generating $N_2 - N_1 + 1$ orders of equal intensity in the interval $[N_1\Delta, N_2\Delta]$, where $\Delta = \lambda z_1/2\pi$

$$N_1 = -\text{int}\left[\frac{R^2}{2\lambda K z_2}\right], \quad N_2 = -\,\text{int}\left[\frac{R^2}{2\lambda K z_1}\right] \tag{7.71}$$

Then, the general equation (6.61) yields coordinates of the foci of the multifocus lens of Eqs (7.68)–(7.70) in the form

$$F_j = -\frac{R^2}{2\lambda Kj}, \quad j = \overline{N_1, N_2} \tag{7.72}$$

where N_1 and N_2 are specified by Eq. (7.71).

Thus, for $\tilde{\varphi}(r) \equiv 0$, the intensity distribution on the optical axis formed by the quasi-periodic DOE of Eq. (7.66) corresponds to N peaks, with the neighbouring peaks' width K times less than the distance between neighbouring foci. For $K = 1$, the peaks merge, producing a line segment along the optical axis. The situation is equivalent to the fact that the diffraction grating ensures disintegration of a focal pattern into a set of spots only for the number of periods $K > 1$. The $\tilde{\varphi}(r)$ function of Eq. (7.65) corresponds to the spherical aberration pixels in Eq. (7.61) upon focusing into a segment $[F_j, F_{j+1}]$. This provides the formation of a continuous distribution along the optical axis at $K > 1$.

Note that the foregoing analysis is only a qualitative explanation of a quasi-periodic DOE of Eq. (7.66) operation. A rigorous substantiation of the form of the function $\tilde{\varphi}(r)$ follows from Eq. (7.58) and from the technique used for calculating a quasi-periodic DOE [79] focusing into a transverse line segment.

7.4.2 Calculation of Binary Plates

For $L/z_2 \ll 1$, the calculation of a DOE intended for focusing the plane beam into the axial segment $z \in [z_1, z_2]$ with uniform intensity reduces to the calculation of the phase function $\varphi_1(x), x \in [0, R^2/2]$ of a 1D DOE focusing the convergent cylindrical beam into a transverse line segment $z \in [-1, -z_1/z_2]$ with uniform intensity. The $\varphi_1(x)$ ensures focusing into a line segment of length $N\Delta$, where

$$N = \text{int}\left[\frac{LR^2}{2\lambda z_1 z_2}\right] \tag{7.73}$$

and $\Delta = \lambda z_1/(R^2/2)$ is the diffraction size. Assume that $\varphi_b(x), x \in [0, R^2/2]$, denotes the phase function of a binary grating period forming N diffraction orders of equal intensity. The $\varphi_b(x)$ function can be considered as the phase function of a 1D DOE focusing into the line segment $[-N\Delta/2, N\Delta/2]$. Accordingly, we can represent the phase function of the 1D DOE focusing into a line segment $z \in [-1, -z_1/z_2]$ in the form

$$\varphi(x) = -\frac{kx}{2f_1} + \varphi_b(x) \tag{7.74}$$

where

$$f_1 = \frac{2z_1 z_2}{z_2 + z_1}$$

According to Eq. (7.57), the phase function of the DOE aimed at focusing the plane beam into an axial line segment is given by

$$\varphi(r) = -\frac{kr^2}{2f_1} + \Phi_b[r] \tag{7.75}$$

where

$$\Phi_b[r] = \varphi_b(r^2/2) \tag{7.76}$$

The phase function of Eq. (7.75) is represented as a superposition of the phase function of a lens and a binary plate $\Phi_b[r]$.

An analytical solution of Eqs (5.1), (5.2), (5.5), and (5.6) for the N-order binary grating considered in Chapter 5 makes it possible to deduce the phase function $\Phi_b[r]$ in analytical form

$$\Phi_b[r] = \Phi_2\left[\mathrm{mod}_{2\pi}\left(\frac{kLr^4}{8z_1 z_2 R^2}\right)\right] \tag{7.77}$$

where

$$\Phi_2[\xi] = \begin{cases} 0, & \xi \in [0, \pi] \\ \pi, & \xi \in [\pi, 2\pi] \end{cases}$$

It is noteworthy that binary elements are easy to manufacture. For the visible and near-IR band, the generation of a multi-level diffraction microrelief is a challenging problem. In this case, it is found to be expedient, for focusing into an axial line segment, to use a binary plate $\Phi_b[r]$ placed right up to a thin lens of focus f_1.

7.4.3 Results of Numerical Simulation

In this section we numerically study the obtained solutions of the problem of focusing into an axial segment. The considered methods for calculating DOEs are aimed at generating a desired intensity distribution on an optical axis, but do not control the energy

$$E(z, \varepsilon) = 2\pi \int_0^{\varepsilon} I(\rho, z)\rho \mathrm{d}\rho$$

coming to an ε-vicinity of the optical axis. The axial energy distribution $E(z, \varepsilon), z \in [z_1, z_2]$ appears to be the most important in terms of the DOE's practical use. In this connection, to characterize the DOE performance we shall employ the following quantities of the r.m.s. deviation δ and the energy efficiency E. The value

$$\delta(\varepsilon) = \frac{1}{\bar{E}}\left[\frac{1}{(z_2 - z_1)}\int_{z_1}^{z_2} [E(z,\varepsilon) - \bar{E}(\varepsilon)]^2 \mathrm{d}z\right]^{1/2}$$

where

$$\bar{E}(\varepsilon) = \frac{1}{(z_2 - z_1)}\int_{z_1}^{z_2} E(z,\varepsilon)\mathrm{d}z$$

characterizes the r.m.s. deviation of the axial energy distribution $E(z,\varepsilon)$ from an average value $\bar{E}(\varepsilon)$. If $\varepsilon \to 0$, $\delta(\varepsilon)$ characterizes the r.m.s. deviation of the optical axis intensity distribution from the average value.

The energy efficiency

$$E(\varepsilon) = \bar{E}(\varepsilon) \Bigg/ \left(2\pi\int_0^R I_0(r)r\mathrm{d}r\right)$$

gives an average portion of illuminating beam energy focused into a cross-section of the focusing segment of radius ε.

Table 7.7 gives the calculated values of $E(\varepsilon)$ and $\delta(\varepsilon)$ for a focusator with the phase function of Eq. (7.60) intended for focusing a plane beam into an axial segment with unform intensity for the following parameters: $\lambda = 0.63\ \mu$m, $z_1 = 320$ mm, $z_2 = 360$ mm, and $R = 15$ mm. It also gives similar values for an 'iterative DOE'. For the iterative DOE, the $\varphi_1(x)$ function was calculated via the iterative AA algorithm described in Chapter 1. The quantity $\Delta = 0.61\lambda(z_1 + z_2)/(2R)$ in the first column corresponds to the spot-size for a lens with focus $f = (z_1 + z_2)/2$. Table 7.7 shows that a focusator having an axial nonuniformity of intensity distribution equal to 29% is characterized by a heavy nonuniformity of the energy distribution $E(z,\varepsilon)$ varying from 78% at $\varepsilon = \Delta$ to 59% at $\varepsilon = 3\Delta$. For the iterative DOE, the axial nonuniformity of intensity distribution decreases by a factor of more than 4, at considerably smaller decrement (~1.15 times) of the r.m.s. error $\delta(\varepsilon)$ at $\varepsilon = \Delta \div 3\Delta$.

Table 7.7 Energy efficiency $E(\varepsilon)$ and r.m.s. deviation $\delta(\varepsilon)$ for a focusator and for an iterative DOE

	Focusator of Eq. (7.60)		Iterative DOE	
ε	$\delta(\varepsilon)$ (%)	$E(\varepsilon)$ (%)	$\delta(\varepsilon)$ (%)	$E(\varepsilon)$ (%)
$\varepsilon \ll \Delta$	29.1	—	7.1	—
Δ	77.8	1.8	63.4	1.9
2Δ	65.6	3.6	56.8	3.6
3Δ	59.3	5.3	49.4	5.1

Table 7.8 Energy efficiency $E(\varepsilon)$ and r.m.s. deviation $\delta(\varepsilon)$ for quasi-periodic DOEs

	Quasi-periodic DOE of Eq. (7.64)				Iterative quasi-periodic DOE			
	$K=2$		$K=4$		$K=2$		$K=4$	
ε	$\delta(\varepsilon)$ (%)	$E(\varepsilon)$ (%)	$\delta(\varepsilon)$ (%)	$E(\varepsilon)$ (%)	$\delta(\varepsilon)$ (%)	$E(\varepsilon)$ (%)	$\delta(\varepsilon)$ (%)	$E(\varepsilon)$ (%)
$\varepsilon \ll \Delta$	31.0	—	35.3	—	12.5	—	11.1	—
Δ	50.1	1.9	49.9	1.8	38.9	1.7	38.0	1.8
2Δ	50.8	3.7	53.3	3.7	44.6	3.3	42.9	3.6
3Δ	41.8	5.4	40.3	5.4	36.5	4.8	32.2	5.3

Table 7.8 gives values of $E(\varepsilon)$ and $\delta(\varepsilon)$ for quasi-periodic DOEs with the phase function of Eq. (7.64) and for an 'iterative quasi-periodic DOE', at $K = 2, 4$. For iterative quasi-periodic DOEs, the functions $\varphi_p(x)$ in Eq. (7.63) were calculated via an iterative AA algorithm. An analysis of Tables 7.7 and 7.8 shows that for quasi-periodic DOEs, the r.m.s. error of axial intensity distribution is greater than that for the focusator and amounts to 31%, for $K = 2$, and 35%, for $K = 4$. Nonuniformity of the axial intensity distribution for quasi-periodic DOEs increases due to the increase of ray-tracing solution errors with increasing K. At the same time, the energy distribution $E(z, \varepsilon)$ nonuniformity (at $\varepsilon = \Delta \div 3\Delta$) for quasi-periodic DOEs is 1.4–1.5 times less than that for focusators. The r.m.s. error $\delta(\varepsilon)$ at $\varepsilon = \Delta \div 3\Delta$ for quasi-periodic DOEs is ~1.2 times less than that for iterative DOEs. For the iterative quasi-periodic DOEs, as compared with the quasi-periodic DOEs of Eq. (7.64), the axial nonuniformity of intensity distribution decreases ~3 times, at a lesser decrement (1.2 to 1.3 times) of the r.m.s. error $\delta(\varepsilon)$ at $\varepsilon = \Delta \div 3\Delta$. The energy efficiency $E(\varepsilon)$ is approximately equivalent for all DOEs studied and varies from 1.7–1.8% at $\varepsilon = \Delta$ to 5.1–5.3% at $\varepsilon = 3\Delta$.

Table 7.9 gives values of $E(\varepsilon)$ and $\delta(\varepsilon)$ for the binary plate of Eq. (7.77)

Table 7.9 Energy efficiency $E(\varepsilon)$ and r.m.s. deviation $\delta(\varepsilon)$ for binary plates

	Binary plate of Eq. (7.77)		Iterative binary plate	
ε	$\delta(\varepsilon)$ (%)	$E(\varepsilon)$ (%)	$\delta(\varepsilon)$ (%)	$E(\varepsilon)$ (%)
$\varepsilon \ll \Delta$	33.9	—	8.8	—
Δ	81.0	1.8	74.1	1.7
2Δ	65.6	3.7	68.5	3.5
3Δ	56.8	5.4	55.0	5.1

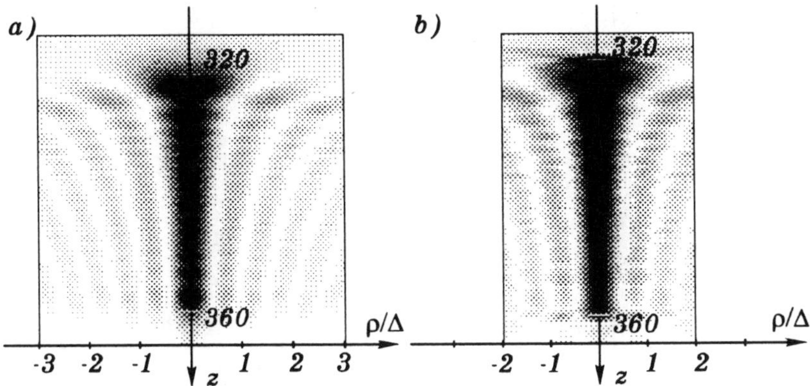

Figure 7.20 Grade-level intensity distributions in the (ρ, z) plane: (a) for the focusator of Eq. (7.60); (b) for the iterative DOE

and for the 'iterative binary plate'. For the iterative binary plate, the $\varphi_b(x)$ function in Eq. (7.76) was calculated via a gradient method, as described in Chapter 5. An analysis of Tables 7.7 and 7.9 shows that the difference of the r.m.s. errors for the binary plate of Eq. (7.77) and for the focusator of Eq. (7.60) is only 4–6%. For the iterative binary plate, as compared with the binary plate of Eq. (7.77), the axial nonuniformity of intensity distribution is 3.5 times less, at a lesser decrement (7–1.5%) of the r.m.s. error $\delta(\varepsilon)$ at $\varepsilon = \Delta \div 3\Delta$.

Figure 7.20 illustrates the calculated grade-level intensity distributions $I(\rho, z)$ for the focusator and for the iterative DOE, respectively. In Fig. 7.20a one can see a pronounced diffusion of the field at the segment beginning, with the energy basically concentrated within a focal spot with a width ranging from $1.5\Delta \div 2\Delta$ at the segment beginning to Δ at the segment end. The ray-tracing method used for the calculation of the $\varphi_1(x)$ function allows us to treat the focusator focusing into an axial segment as a set of concentric rings, with each of them forming part of an axial segment. The central ring contributes to the beginning of the segment, whereas external rings focus at the segment's end. The spot-sizes for the internal rings are greater than the spot-sizes for the external ones; this explains the decrement of the field diffusion to the segment end. For the iterative DOE, the nonuniformity of the axial intensity distribution is 4 times less, which, however, does not compensate for the pronounced field diffusion at the segment beginning (see Fig. 7.20b).

Grade-level intensity distributions $I(\rho, z)$ for the quasi-periodic DOE of Eq. (7.64) and for the iterative quasi-periodic DOE at $K = 4$ are shown in Fig. 7.21. The fields in Fig. 7.21 are periodic in structure. The results reported are consistent with the explanation of the quasi-periodic DOE operation given in section 7.4.1. At $\tilde{\varphi}(r) \equiv 0$, the quasi-periodic DOE of Eq. (7.64)

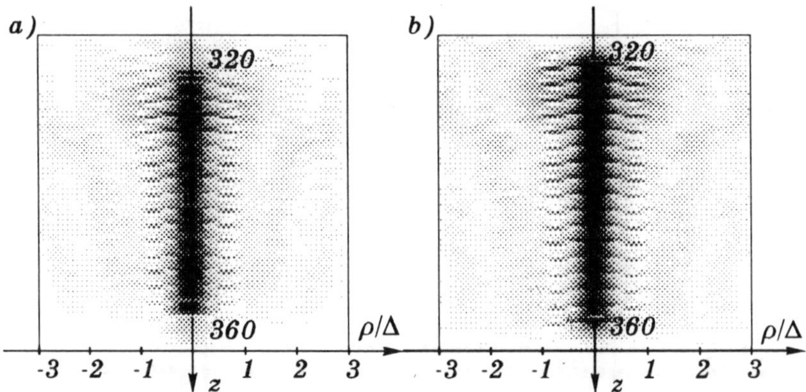

Figure 7.21 Grade-level intensity distribution in the (ρ, z) plane: (a) for the quasi-periodic DOE of Eq. (7.64); (b) for the iterative quasi-periodic DOE

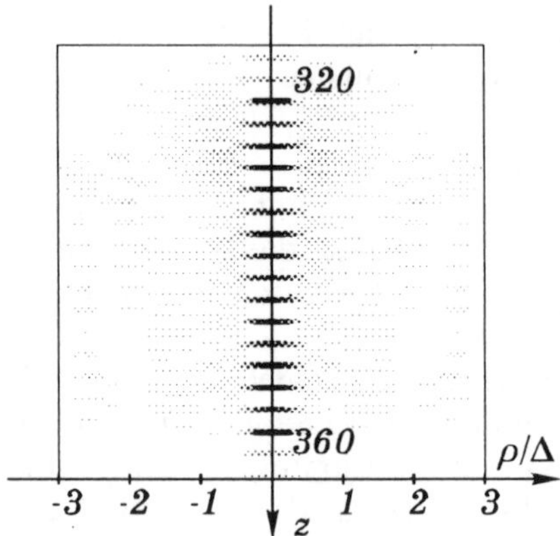

Figure 7.22 Grade-level intensity distribution in the (ρ, z) plane for the quasi-periodic DOE of Eq. (7.64) at $\tilde{\varphi}(r) \equiv 0$

is equivalent to a multifocal lens (see Fig. 7.22). The $\tilde{\varphi}(r)$ function of Eq. (7.65) corresponds to the pixels of spherical aberration and provides the formation of an interfocal field. The number N of the field periods is an inverse proportion to K and is determined by Eq. (7.67) (for the above parameters $N = 15$). Thus, the value of K affects the field structure formed by the quasi-periodic DOE. We can then consider the K value as an additional optimization parameter in calculating the DOE, making it possible to optimize the values of the energy efficiency and the r.m.s. error. For example, the iterative quasi-periodic DOE has a higher efficiency and forms more uniform energy distributions for $K = 4$ than for $K = 2$ (see Table 7.8).

Figure 7.23 illustrates the calculated intensity distributions $I(\rho, z)$ for the

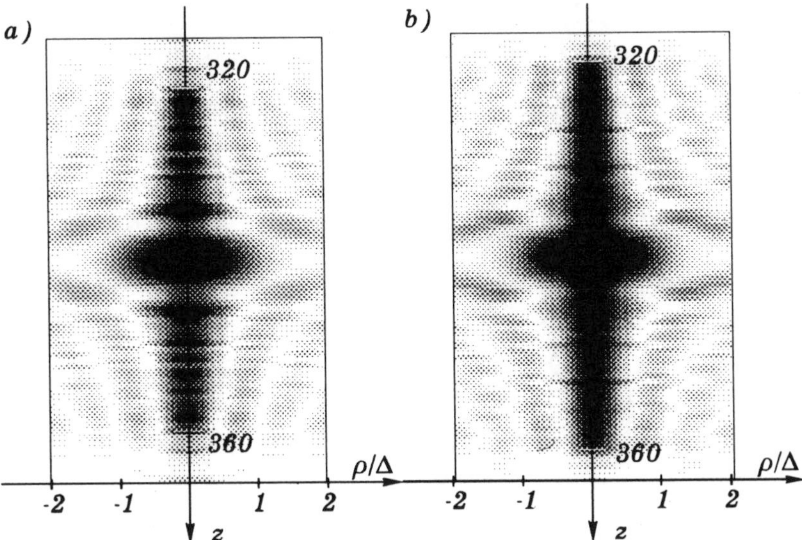

Figure 7.23 Grade-level intensity distribution in the (ρ, z) plane: (a) for the binary plate of Eq. (7.77); (b) for the iterative binary plate of Eq. (7.76)

binary plate of Eq. (7.77) and for an iterative binary plate coupled with a thin lens. In Fig. 7.23a the pronounced field diffusion occurs at the segment centre since the central part of the binary plate contributes to the segment centre. The external parts focus at the ends of the segment, thus decreasing the field diffusion at the segment ends. The iterative binary plate evens an axial intensity distribution but does not compensate for the pronounced field diffusion at the segment centre (see Fig. 7.23b).

The numerical investigation makes it possible to choose a method for calculating the DOE with regard to requirements as to the field distribution along the optical axis in a particular task.

Conclusion

It is clear that the algorithms and methods dealt with in this book do not cover either the entire spectrum of iterative methods employed in optics or the entire class of problems related to the computation of the phase of diffractive optical elements. However, the algorithms reported in this work offer practical, not just mathematical, advantages. We briefly outline the tasks in which iterative computation enables one to obtain new properties of optical elements, in comparison with non-iterative computation.

1. For the DOEs that are called focusators and employed to focus the high energy laser light into curvilinear line segments, with a minimum diffraction width, the iterative algorithms of synthesis make it possible to remove the limitations on the energy distribution along the line and take account of diffraction effects on the edges of the segments.

2. For the phase diffraction gratings used as multichannel illuminators in parallel data processing, the use of iterative methods enables an essential enhancement of the number of diffraction orders (128×128 and more), with the energy distribution controlled between the orders.

3. For the DOEs that produce the light modes (Bessel, Gauss–Laguerre, Gauss–Hermite) propagating in different diffraction orders and which have found use for the simultaneous input of laser light into a set of fibres, the iterative methods surpass the coding methods of digital holography and offer energy efficiency close to 100%.

4. A purposeful introduction of nonlinear phase predistortions enabled the design of multiorder DOEs forming the desired image on desired planes along the axis of light propagation. An example of such an element is multifocus lenses. In this case, the iterative methods are employed in the design of both the basic phase and the predistorting function.

5. Appendices F and G reflect an attempt to employ the iterative

algorithms for the synthesis of the surface relief of the DOE with consideration for light polarization. The authors hope that the iterative (gradient) methods will enable an advance in solving the task of DOE synthesis using the electromagnetic approximation.

6. Some of the above optical elements for the visible and IR range designed via the iterative methods were fabricated microlithographically and tested in laboratory conditions. Note, however, that the experimental and technological aspects of DOE synthesis are beyond the scope of this book.

7. The iterative algorithms permit a simple software realization: all the programs are in the C++ language and all the calculations are made using the IBM PC/AT-286.

Appendix A

Proof of Convergence of the Adaptive–Additive Algorithm

To prove that the AA algorithm of Eq. (1.19) or (1.40) converges on average, we introduce the following designations

$$W_n(u,v) = A_0 e^{i\varphi_n}, \quad \overline{W}_n(u,v) = A_n e^{i\varphi_n} \tag{A1}$$

$$F_n(\xi,\eta) = B_n e^{i\psi_n}, \quad \overline{F}_n(\xi,\eta) = \overline{B}_n e^{i\psi_n} \tag{A2}$$

$$\overline{B}_n = \lambda B_0 + (1-\lambda) B_n \tag{A3}$$

Note that here $\overline{W}_n$ and W_n denote, respectively, light complex amplitudes in the DOE plane before and after replacement and are related to the variables $\overline{F}_n$ and F_n, denoting complex amplitudes in the Fourier-image plane at the input and output, via

$$B_n e^{i\psi_n} = \mathfrak{F}[A_0 e^{i\varphi_n}], \quad A_{n+1} e^{i\varphi_{n+1}} = \mathfrak{F}^{-1}[\overline{B}_n e^{i\varphi_n}] \tag{A4}$$

where $\mathfrak{F}$ and $\mathfrak{F}^{-1}$ are the direct and inverse Fourier transforms.

Let us show then that for any number of iterations the following inequality holds

$$\iint_{-\infty}^{\infty} |B_{n+1} - B_0|^2 \mathrm{d}\xi \mathrm{d}\eta \leq \iint_{-\infty}^{\infty} |B_n - B_0|^2 \mathrm{d}\xi \mathrm{d}\eta \tag{A5}$$

Actually, it appears to be easy to demonstrate the validity of the following chain of equalities

$$\iint_{-\infty}^{\infty} |\overline{B}_{n+1} - B_{n+1}|^2 \mathrm{d}\xi \mathrm{d}\eta = \iint_{-\infty}^{\infty} |\lambda B_0 + (1-\lambda) B_{n+1} - B_{n+1}|^2 \mathrm{d}\xi \mathrm{d}\eta \tag{A6}$$

$$= \lambda^2 \iint_{-\infty}^{\infty} |B_0 - B_{n+1}|^2 \mathrm{d}\xi \mathrm{d}\eta = \lambda^2 \iint_{-\infty}^{\infty} |B_0 e^{i\psi_{n+1}} - B_{n+1} e^{i\psi_{n+1}}|^2 \mathrm{d}\xi \mathrm{d}\eta$$

Based on the triangle inequality

$$|a| - |b| \leq |a - b| \tag{A7}$$

which holds for any complex numbers a and b we obtain the inequality

$$\iint_{-\infty}^{\infty} |B_0 e^{i\psi_{n+1}} - B_{n+1} e^{i\psi_{n+1}}|^2 d\xi d\eta \leq \iint_{-\infty}^{\infty} |B_0 e^{i\psi_n} - B_{n+1} e^{i\psi_{n+1}}|^2 d\xi d\eta \tag{A8}$$

Extending the chain of Eqs (A6) and (A8) and using the Parseval equality

$$\iint_{-\infty}^{\infty} |B_n(\xi, \eta)|^2 d\xi d\eta = \iint_{-\infty}^{\infty} |A_0(x, y)|^2 dx\, dy \tag{A9}$$

yields

$$\begin{aligned} &\iint_{-\infty}^{\infty} |\lambda B_0 e^{i\psi_n} - \lambda B_{n+1} e^{i\psi_{n+1}}|^2 d\xi d\eta \\ &= \iint_{-\infty}^{\infty} |\lambda B_0 e^{i\psi_n} + (1-\lambda) B_n e^{i\psi_n} \\ &- (1-\lambda) B_n e^{i\psi_n} - \lambda B_{n+1} e^{i\psi_{n+1}}|^2 d\xi d\eta \\ &= \iint_{-\infty}^{\infty} |A_{n+1} e^{i\varphi_{n+1}} - (1-\lambda) A_0 e^{i\varphi_n} - \lambda A_0 e^{i\varphi_{n+1}}|^2 du\, dv \end{aligned} \tag{A10}$$

By demonstrating that the following inequality is observed

$$\begin{aligned} &\iint_{-\infty}^{\infty} |A_{n+1} e^{i\varphi_{n+1}} - (1-\lambda) A_0 e^{i\varphi_n} - \lambda A_0 e^{i\varphi_{n+1}}|^2 du\, dv \\ &\leq \iint_{-\infty}^{\infty} |A_{n+1} e^{i\varphi_{n+1}} - (1-\lambda) A_0 e^{i\varphi_n} - \lambda A_0 e^{i\varphi_n}|^2 du\, dv \end{aligned} \tag{A11}$$

we can continue the chain of Eqs (A6), (A8), and (A10) as follows

$$\begin{aligned} &\iint_{-\infty}^{\infty} |A_{n+1} e^{i\varphi_{n+1}} - (1-\lambda) A_0 e^{i\varphi_n} - \lambda A_0 e^{i\varphi_n}|^2 dx\, dy \\ &= \iint_{-\infty}^{\infty} |A_{n+1} e^{i\varphi_{n+1}} - A_0 e^{i\varphi_n}|^2 dx\, dy \\ &= \iint_{-\infty}^{\infty} |[\lambda B_0 + (1-\lambda) B_n] e^{i\psi_n} - B_n e^{i\psi_n}|^2 d\xi d\eta \\ &= \iint_{-\infty}^{\infty} |\lambda B_0 e^{i\psi_n} - \lambda B_n e^{i\psi_n}|^2 d\xi d\eta \\ &= \lambda^2 \iint_{-\infty}^{\infty} |B_0 - B_n|^2 d\xi d\eta \end{aligned} \tag{A12}$$

The second equality in Eq. (A12) follows from the Parseval equality; the others are obvious. From the chain of Eqs (A6), (A8), and (A10)–(A12) it follows that inequality (A5) is valid.

Let us demonstrate the truth of inequality (A11). Consider a function $I(\lambda)$ of λ in the form

$$I(\lambda) = |(A_{n+1} - \lambda A_0)e^{i\varphi_{n+1}} - (1-\lambda)A_0 e^{i\varphi_n}|^2 \tag{A13}$$

which is an integrand of the first integral in Eq. (A11). The function of Eq. (A13) can also be represented in the canonical form of the quadratic parabola

$$I(\lambda) = p\lambda^2 - q\lambda + I(0) \tag{A14}$$

where

$$p = 4A_0^2 \sin^2\left[\frac{\varphi_{n+1} - \varphi_n}{2}\right] \tag{A15}$$

$$q = 4A_0(A_0 + A_{n+1}) \sin^2\left[\frac{\varphi_{n+1} - \varphi_n}{2}\right] \tag{A16}$$

$$I(0) = |A_{n+1}e^{i\varphi_{n+1}} - A_0 e^{i\varphi_n}|^2 \tag{A17}$$

Note that the quantity $I(0)$ is an integrand of the second integral in Eq. (A11).

Because p is positive, the parabola specified by Eq. (A14) will have a minimum at point

$$\lambda_0 = \frac{q}{2p} = \frac{A_0 + A_{n+1}}{2A_0} \tag{A18}$$

and the inequality

$$I(\lambda) \leq I(0) \tag{A19}$$

will hold under the condition

$$0 \leq \lambda \leq 2\lambda_0 = 1 + \frac{A_{n+1}}{A_0} \tag{A20}$$

The fulfilment of inequality (A19) guarantees the fulfilment of inequality (A11). Inequality (A20) determines the strategy of varying the λ parameter in the course of iterations. At the beginning of the iterative process, when A_n differs greatly from A_0, the λ parameter should be chosen from the interval $[0, 1]$. After several iterations the value of A_n will approach A_0, and the λ parameter can be chosen from the interval $[1, 2]$.

Appendix B

Proof of Convergence of the Algorithm for Computing Phase Formers of Wavefronts

The proof of the truth of inequality (3.5) guaranteeing the convergence of the iterative procedure given by Eqs (3.1)–(3.3) has been deduced under the condition there is only a small difference between the complex amplitudes in the observation plane for two adjacent iterations.

Let us introduce the following notation for the complex amplitudes in the nth step

$$F_n(\xi, \eta) = B_n \mathrm{e}^{i\psi_n} \tag{B1}$$

$$W_n(u, v) = A_n \mathrm{e}^{i\varphi_n} \tag{B2}$$

and notation for the relation between the amplitudes in different planes

$$B_n \mathrm{e}^{i\psi_n} = FR[A_0 \mathrm{e}^{i\varphi_n}] \tag{B3}$$

$$A_{n+1} \mathrm{e}^{i\varphi_{n+1}} = FR^{-1}[B_n \mathrm{e}^{i\psi_0}] \tag{B4}$$

where A_0 is a desired amplitude in the kinoform plane, ψ_0 is a desired phase in the observation plane, and FR and FR^{-1} denote the direct and inverse Fresnel transforms.

It is easy to show that the Parseval equality holds, which means the conservation of the total light energy on the transverse planes

$$\int\int_{-\infty}^{\infty} |A_0 \mathrm{e}^{i\varphi n} \mathrm{d}u \mathrm{d}v = \int\int_{-\infty}^{\infty} |B_n \mathrm{e}^{i\psi_n}|^2 \mathrm{d}\xi \mathrm{d}\eta \tag{B5}$$

An obvious property of complex numbers i.e. that the difference of two complex numbers with different arguments (less than π) is greater than the difference of two complex numbers with equal arguments, results in the inequality

$$\begin{aligned}\iint_{-\infty}^{\infty} |A_{n+1} - A_0|^2 \mathrm{d}u\,\mathrm{d}v &= \iint_{-\infty}^{\infty} |A_{n+1}\mathrm{e}^{i\varphi_{n+1}} - A_0\mathrm{e}^{i\varphi_{n+1}}|^2 \mathrm{d}u\,\mathrm{d}v \\ &\leq \iint_{-\infty}^{\infty} |A_{n+1}\mathrm{e}^{i\varphi_{n+1}} - A_0\mathrm{e}^{i\varphi_n}|^2 \mathrm{d}u\,\mathrm{d}v\end{aligned} \tag{B6}$$

Based on the Parseval equality of Eq. (B5), we can write the relation

$$\iint_{-\infty}^{\infty} |A_{n+1}\mathrm{e}^{i\varphi_{n+1}} - A_0\mathrm{e}^{i\varphi_n}|^2 \mathrm{d}u\,\mathrm{d}v = \iint_{-\infty}^{\infty} |B_n\mathrm{e}^{i\psi_0} - B_n\mathrm{e}^{i\psi_n}|^2 \mathrm{d}\xi\,\mathrm{d}\eta \tag{B7}$$

Let us show then that the following inequality holds 'in the small'

$$\iint_{-\infty}^{\infty} |B_n\mathrm{e}^{i\psi_0} - B_n\mathrm{e}^{i\psi_n}|^2 \mathrm{d}\xi\,\mathrm{d}\eta \leq \iint_{-\infty}^{\infty} |B_{n-1}\mathrm{e}^{i\psi_0} - B_n\mathrm{e}^{i\psi_n}|^2 \mathrm{d}\xi\,\mathrm{d}\eta \tag{B8}$$

Actually, given that

$$\psi_n - \psi_0 = \Delta\psi, \quad |\Delta\psi| \ll 1 \tag{B9}$$

$$\frac{B_{n-1}}{B_n} = 1 + \Delta B, \quad |\Delta B| \ll 1 \tag{B10}$$

we find that

$$|\mathrm{e}^{i\psi_n} - \mathrm{e}^{i\psi_0}|^2 = 4\sin^2\left[\frac{\psi_n - \psi_0}{2}\right] \approx \Delta\psi^2 \tag{B11}$$

$$\begin{aligned}&\left|\mathrm{e}^{i\psi_n} - \frac{B_{n-1}}{B_n}\mathrm{e}^{i\psi_0}\right|^2 \\ &= 1 + \left[\frac{B_{n-1}}{B_n}\right]^2 - 2\left[\frac{B_{n-1}}{B_n}\right]\cos(\psi_n - \psi_0) \approx \Delta B^2 + \Delta\psi^2\end{aligned} \tag{B12}$$

From formulae (B9) to (B12) it follows from (B8) that

$$\begin{aligned}\iint_{-\infty}^{\infty} |B_n\mathrm{e}^{i\psi_0} - B_n\mathrm{e}^{i\psi_n}|^2 \mathrm{d}\xi\,\mathrm{d}\eta &= \iint_{-\infty}^{\infty} B_n^2|\mathrm{e}^{i\psi_0} - \mathrm{e}^{i\psi_n}|^2 \mathrm{d}\xi\,\mathrm{d}\eta \\ &\approx \iint_{-\infty}^{\infty} B_n^2\Delta\psi^2 \mathrm{d}\xi\,\mathrm{d}\eta \leq \iint_{-\infty}^{\infty} B_n^2[\Delta B^2 + \Delta\psi^2] \mathrm{d}\xi\,\mathrm{d}\eta \\ &\approx \iint_{-\infty}^{\infty} |B_n\mathrm{e}^{i\psi_n} - B_{n-1}\mathrm{e}^{i\psi_0}|^2 \mathrm{d}\xi\,\mathrm{d}\eta\end{aligned} \tag{B13}$$

Next, the repeated employment of the Parseval equality (B5) gives

$$\int\int_{-\infty}^{\infty} |B_{n-1}e^{i\psi_0} - B_n e^{i\psi_n}|^2 d\xi d\eta = \int\int_{-\infty}^{\infty} |A_n e^{i\varphi_n} - A_0 e^{i\varphi_n}|^2 du\, dv$$
$$= \int\int_{-\infty}^{\infty} |A_n - A_0|^2 du dv \tag{B14}$$

Finally, relations (B6) to (B8) and (B14) give

$$\int\int_{-\infty}^{\infty} |A_{n+1} - A_0|^2 du\, dv \leq \int\int_{-\infty}^{\infty} |A_n - A_0|^2 du\, dv \tag{B15}$$

Inequality (B15) testifies that the iterative procedure of the calculation of the phase distribution at a certain distance converges if two successive estimates of the light complex amplitude in the observation plane are taken to be close to each other. However, in practice, as is seen from numerical examples, this algorithm converges at any initial estimate of the light complex amplitude.

Appendix C

Proof of Convergence of the Algorithm for Calculating Amplitude Transparencies for the Formation of Wavefronts

To prove the inequality (3.12), let us introduce the designations

$$B_n e^{i\psi_n} = FR[A_{n-1}e^{i\varphi_0}] \tag{C1}$$

$$A_n e^{i\varphi_n} = FR^{-1}[B_n e^{i\psi_0}] \tag{C2}$$

where B and A are the amplitudes in the planes of observation and transparency, ψ and φ are the phases in the planes of observation and transparency, respectively, FR and FR^{-1} denote the direct and inverse Fresnel transform, and n is the number of iterations.

From the condition of conservation of light energy we can derive the Parseval equation (B5) from which, in combination with the statement that the module of difference of two complex numbers with different arguments is greater than the difference of two complex numbers with equal arguments, we obtain the chain of inequalities

$$\begin{aligned}
\iint_{-\infty}^{\infty} |B_{n+1} - B_n|^2 \mathrm{d}\xi \mathrm{d}\eta &= \iint_{-\infty}^{\infty} |B_{n+1}e^{i\psi_{n+1}} - B_n e^{i\psi_{n+1}}|^2 \mathrm{d}\xi \mathrm{d}\eta \\
\leq \iint_{-\infty}^{\infty} |B_{n+1}e^{i\psi_{n+1}} - B_n e^{i\psi_n}|^2 \mathrm{d}\xi \mathrm{d}\eta &= \iint_{-\infty}^{\infty} |A_n e^{i\varphi_0} - A_{n-1}e^{i\varphi_0}|^2 \mathrm{d}u\, \mathrm{d}v \\
\leq \iint_{-\infty}^{\infty} |A_n e^{i\varphi_n} - A_{n-1}e^{i\varphi_{n-1}}|^2 \mathrm{d}u\, \mathrm{d}v &= \iint_{-\infty}^{\infty} |B_n e^{i\psi_0} - B_{n-1}e^{i\psi_0}|^2 \mathrm{d}\xi \mathrm{d}\eta \\
&= \iint_{-\infty}^{\infty} |B_n - B_{n-1}|^2 \mathrm{d}\xi \mathrm{d}\eta
\end{aligned} \tag{C3}$$

Comparison of the first and last expressions in the chain yields the final inequality

$$\int\int_{-\infty}^{\infty} |B_{n+1} - B_n|^2 \mathrm{d}\xi \mathrm{d}\eta \leq \int\int_{-\infty}^{\infty} |B_n - B_{n-1}|^2 \mathrm{d}\xi \mathrm{d}\eta \qquad \text{(C4)}$$

which shows that the r.m.s. error for two successive estimates of the light field amplitude in the observation plane shows no increase in the course of iterations.

Appendix D

Proof of Convergence of the Algorithm for Computing Formers of Bessel Modes

Denoting the basis function

$$\Omega_{nm}(r,\varphi) = A_{nm} J_n\left(\gamma_{nm}\frac{r}{R}\right) e^{-in\varphi} \tag{D1}$$

where

$$A_{nm} = \{\pi[RJ'_n(\gamma_{nm})]^2\}^{-1/2}$$
$$J_{-n}(x) = (-1)^n J_n(x), \quad \gamma_{(-n)m} = \gamma_{nm}$$

are the roots of the Bessel function of the nth order, and $J_n(\gamma_{nm}) = 0$.

Any continuous function limited in a circle of radius R can be expanded into a series in terms of the basis in Eq. (D1)

$$U(r,\varphi) = \sum_{n=-\infty}^{\infty}\sum_{m=0}^{\infty} C_{nm}\Omega_{nm}(r,\varphi) \tag{D2}$$

The orthogonality condition for the function in Eq. (D1) takes the form

$$\int_0^R\int_0^{2\pi}\Omega_{nm}(r,\varphi)\,\Omega^*_{kl}(r,\varphi)\,r\,\mathrm{d}r\,\mathrm{d}\varphi = \delta_{nk}\delta_{ml} \tag{D3}$$

Let $U_{p+1}(r,\varphi)$ denote the estimate of the function $U(r,\varphi)$ at the $(p+1)$th iterative step and let it be found from the equation

$$\begin{aligned} U_{p+1}(r,\varphi) &= W_{p+1}(r,\varphi)\exp[i\Psi_{p+1}(r,\varphi)] \\ &= \sum_{n=-\infty}^{\infty} B_n \exp[iv_n^{(p)}]\,\Omega_n(r,\varphi) \end{aligned} \tag{D4}$$

where

$$W_{p+1}(r,\varphi)=|U_{p+1}(r,\varphi)|,\quad \Omega_n=\Omega_{nm}|_{m=m_0}$$

is a subset of the basis functions with fixed m, $v_m^{(p)}=\arg[C_{nm}^{(p)}]|_{m=m_0}$ is a subset of arguments of the coefficients in the expansion (D2) for fixed $m=m_0$, and $B_n\geq 0$ are the pregiven numbers.

From Eq. (D4) follows the Parseval equality

$$\sum_{n=-\infty}^{\infty} B_n^2=\int_0^R\int_0^{2\pi} W_{p+1}^2(r,\varphi)\,r\mathrm{d}r\mathrm{d}\varphi \tag{D5}$$

The coefficients of the function $C_{nm}^{(p)}$ can be derived from the expansion of the function in terms of the basis

$$W_0(r,\varphi)\exp[i\Psi_p(r,\varphi)]=\sum_{n=-\infty}^{\infty}\sum_{m=0}^{\infty}|C_{nm}^{(p)}|\exp\{iv_{nm}^{(p)}\}\,\Omega_{nm}(r,\varphi) \tag{D6}$$

where $W_0(r,\varphi)$ is a pregiven function satisfying the condition of energy conservation

$$\int_0^R W_0^2(r,\varphi)\,r\mathrm{d}r\mathrm{d}\varphi=\int_0^R W_p^2(r,\varphi)\,r\mathrm{d}r\mathrm{d}\varphi \tag{D7}$$

for any number p.

From the expansion in Eq. (D6) and the orthogonality condition in Eq. (D3) follows the equation for calculating the coefficients

$$C_{nm}^{(p)}=\int_0^R\int_0^{2\pi} W_0(r,\varphi)\exp[i\Psi_p(r,\varphi)]\,\Omega_{nm}^*(r,\varphi)\,r\mathrm{d}r\mathrm{d}\varphi \tag{D8}$$

From the expansion in Eq. (D6) also follows the Parseval equality:

$$\sum_{n=-\infty}^{\infty}\sum_{m=0}^{\infty}|C_{nm}^{(p)}|^2=\int_0^R\int_0^{2\pi} W_0^2(r,\varphi)\,r\mathrm{d}r\mathrm{d}\varphi \tag{D9}$$

Using Eqs (D5) and (D9), we can derive the chain of equalities and inequalities

$$\sum_{n=-\infty}^{\infty}\sum_{m=0}^{\infty}[|C_{nm}^{(p+1)}|-B_n]^2=\sum_{n=-\infty}^{\infty}\sum_{m=0}^{\infty}\||C_{nm}^{(p+1)}|\exp\{iv_{nm}^{(p+1)}\}$$

$$-B_n\exp\{iv_{nm}^{(p+1)}\}|^2\leq\sum_{n=-\infty}^{\infty}\sum_{m=0}^{\infty}\||C_{nm}^{(p+1)}|\exp\{iv_{nm}^{(p+1)}\}-B_n\exp\{iv_n^{(p)}\}|^2$$

$$=\int_0^R\int_0^{2\pi}|W_0(r,\varphi)\exp\{i\Psi_{p+1}(r,\varphi)\}-W_{p+1}\exp\{i\Psi_{p+1}(r,\varphi)\}|^2 r\mathrm{d}r\mathrm{d}\varphi \tag{D10}$$

$$\leq\int_0^R\int_0^{2\pi}|W_0(r,\varphi)\exp\{i\Psi_p(r,\varphi)\}-W_{p+1}(r,\varphi)\exp\{i\Psi_{p+1}(r,\varphi)\}|^2 r\mathrm{d}r\mathrm{d}\varphi$$

$$= \sum_{n=-\infty}^{\infty} \sum_{m=0}^{\infty} ||C_{nm}^{(p)}| \exp\{iv_{nm}^{(p)}\} - B_n \exp\{iv_n^{(p)}\}|^2 = \sum_{n=-\infty}^{\infty} \sum_{m=0}^{\infty} ||C_{nm}^{(p)}| - B_n|^2$$

The last equation in relation (D10) follows from

$$B_n \exp\{iv_{nm}^{(p)}\} = B_n \exp\{iv_n^{(p)}\}$$

Note that the inequalities follow from the triangle inequality

$$|\bar{a} - \bar{b}| \geq |\bar{a}| - |\bar{b}| \tag{D11}$$

From (D10) follows an inequality

$$\sum_{n=-\infty}^{\infty} \sum_{m=0}^{\infty} [|C_{nm}^{(p+1)}| - B_n]^2 \leq \sum_{n=-\infty}^{\infty} \sum_{m=0}^{\infty} [|C_{nm}^{(p)}| - B_n]^2 \tag{D12}$$

that can be rearranged as

$$\begin{aligned} &\sum_{n=-\infty}^{\infty} \sum_{m=0}^{\infty} |C_{nm}^{(p+1)}|^2 - 2 \sum_{n=-\infty}^{\infty} |C_n^{(p+1)}| B_n + \sum_{n=-\infty}^{\infty} B_n^2 \\ &\leq \sum_{n=-\infty}^{\infty} \sum_{m=0}^{\infty} |C_{nm}^{(p)}|^2 - 2 \sum_{n=-\infty}^{\infty} |C_n^{(p)}| B_n + \sum_{n=-\infty}^{\infty} B_n^2 \end{aligned} \tag{D13}$$

where $C_n^{(p+1)}$ and $C_n^{(p)}$ are the coefficients C_{nm} of expansion for fixed $m = m_0$ at the $(p+1)$th and pth iterations.

From the condition of energy conservation, Eq. (D9), follows the relation

$$\sum_{n=-\infty}^{\infty} \sum_{m=0}^{\infty} |C_{nm}^{(p+1)}|^2 = \sum_{n=-\infty}^{\infty} \sum_{m=0}^{\infty} |C_{nm}^{(p)}|^2 \tag{D14}$$

In view of Eq. (D14), instead of Eq. (D13) we obtain the inequality

$$\sum_{n=-\infty}^{\infty} |C_n^{(p+1)}| B_n \geq \sum_{n=-\infty}^{\infty} |C_n^{(p)}| B_n \tag{D15}$$

which means that as the number of iterations tends to infinity, the moduli $|C_n^{(p)}|$ will tend to the numbers B_n, on average.

Appendix E

Proof of Convergence of the Algorithm for Calculating Formers of Orthogonal Modes

Since the proof of convergence of the iterative algorithms employing the expansion of the desired complex functions specified by the equations of type (4.26), (4.38), and (4.54) is equivalent, we have been able to present the proof of some abstract task formulated below.

Let us assume that a nonlinear set of algebraic integral equations with respect to the argument of coefficients C_n is to be solved

$$I_0(x, y) = \left| \sum_{n=-\infty}^{\infty} C_n \Psi_n(x, y) \right|^2 \tag{E1}$$

where the functions $\Psi_n(x, y)$ are orthogonal

$$\iint_{-\infty}^{\infty} \Psi_n(x, y) \Psi_m^*(x, y) \, dx \, dy = \delta_{mn} \tag{E2}$$

and the function $I_0(x, y)$ is arbitrary. Assume also the modules of expansion coefficients, $|C_n| = B_n$, to be known and that they are arbitrary positive numbers.

When solving the problem via an iterative approach employing a Gerchberg–Saxton algorithm [14], we should take the following steps.

Assume that in the kth step for the function

$$F(x, y) = \sum_{n=-\infty}^{\infty} C_n \Psi_n(x, y) \tag{E3}$$

we have the estimate

$$\overline{F}_k(x, y) = \surd[I_0(x, y)] \, e^{i\varphi_k(x,y)} \tag{E4}$$

Using the function given by Eq. (E4), we can find the coefficients of the series of Eq. (E3) in the kth iteration step

$$C_n^{(k)} = \iint_{-\infty}^{\infty} \bar{F}_k(x,y)\Psi_n^*(x,y)\,\mathrm{d}x\,\mathrm{d}y \tag{E5}$$

Next, the modules of the coefficients $C_n^{(k)}$ are replaced by the known functions B_n, the phases being unchanged. The next estimate of complex amplitude $F_{k+1}(x,y)$ is deduced from the relation

$$F_{k+1}(x,y) = \sum_{n=-\infty}^{\infty} B_n \exp[i \arg C_n^{(k)}]\Psi_n(x,y) \tag{E6}$$

Therefore, in the $(k+1)$th iteration step the relation given by

$$\varphi_{k+1}(x,y) = \arg F_{k+1}(x,y) \tag{E7}$$

should be substituted in Eq. (E4).

This algorithm for solving the set of Eqs (E1) will converge on average for any number

$$\begin{aligned}&\iint_{-\infty}^{\infty} |\surd[I_0(x,y)] - |F_{k+1}(x,y)||^2\mathrm{d}x\,\mathrm{d}y\\ &\leq \iint_{-\infty}^{\infty} |\surd[I_0(x,y)] - |F_k(x,y)||^2\mathrm{d}x\,\mathrm{d}y\end{aligned} \tag{E8}$$

or

$$\sum_{n=-\infty}^{\infty} |B_n - |C_n^{(k+1)}||^2 \leq \sum_{n=-\infty}^{\infty} |B_n - |C_n^{(k)}||^2 \tag{E9}$$

To prove (E8), let us introduce the designations for the kth iteration step (note that inequality (E9) can be proved similarly)

$$F_k(x,y) = A_k \mathrm{e}^{i\varphi_k},\quad C_n^{(k)} = B_n^{(k)}\mathrm{e}^{i\nu_{nk}},\quad A_0(x,y) = \surd[I_0(x,y)] \tag{E10}$$

$$A_{k+1}\mathrm{e}^{i\varphi_{k+1}} = \sum_{n=-\infty}^{\infty} B_n \mathrm{e}^{i\nu_{nk}}\Psi_n(x,y) \tag{E11}$$

$$A_0\mathrm{e}^{i\varphi_k} = \sum_{n=-\infty}^{\infty} B_n^{(k)}\mathrm{e}^{i\nu_{nk}}\Psi_n(x,y) \tag{E12}$$

Using the orthogonality property given by Eq. (E2) for the function $y_n(x,y)$, we can easily show that the following Parseval-type equality holds

$$\iint_{-\infty}^{\infty} |F_k(x,y)|^2\mathrm{d}x\,\mathrm{d}y = \sum_{n=-\infty}^{\infty} |B_n^{(k)}|^2 \tag{E13}$$

Further, use of Eq. (E13) and the triangle inequality (A7) that has been employed in Appendices A to D gives the chain of inequalities

$$\begin{aligned}\iint_{-\infty}^{\infty}|A_0-A_{k+1}|^2\mathrm{d}x\,\mathrm{d}y &= \iint_{-\infty}^{\infty}|A_0\mathrm{e}^{i\varphi_{k+1}}-A_{k+1}\mathrm{e}^{i\varphi_{k+1}}|^2\mathrm{d}x\,\mathrm{d}y\\ \leq \iint_{-\infty}^{\infty}|A_0\mathrm{e}^{i\varphi_k}-A_{k+1}\mathrm{e}^{i\varphi_{k+1}}|^2\mathrm{d}x\,\mathrm{d}y &= \sum_{n=-\infty}^{\infty}|B_n^{(k)}\mathrm{e}^{i\nu_{nk}}-B_n\mathrm{e}^{i\nu_{nk}}|^2\\ \leq \sum_{n=-\infty}^{\infty}|B_n^{(k)}\mathrm{e}^{i\nu_{nk}}-B_n\mathrm{e}^{i\nu_{nk-1}}|^2 &= \iint_{-\infty}^{\infty}|A_0\mathrm{e}^{i\varphi_k}-A_k\mathrm{e}^{i\varphi_k}|^2\mathrm{d}x\,\mathrm{d}y\\ &= \iint_{-\infty}^{\infty}|A_0-A_k|^2\mathrm{d}x\,\mathrm{d}y\end{aligned} \tag{E14}$$

Comparison of the first and last relations in (E14) yields the final inequality

$$\iint_{-\infty}^{\infty}|A_0-A_{k+1}|^2\mathrm{d}x\,\mathrm{d}y \leq \iint_{-\infty}^{\infty}|A_0-A_k|^2\mathrm{d}x\,\mathrm{d}y \tag{E15}$$

Appendix F

Iterative Design of DOEs Modulating the Coherent Light Polarization

The DOE [1] is a microrelief phase-modulating optical element dedicated to forming a desired intensity distribution (image). The DOE surface microrelief introducing a required phase delay into every point of incident light wave is calculated, as a rule, using iterative methods [2,14]. A microrelief with desired parameters can be produced via microlithographic techniques [4].

The surface microrelief is well suited to governing not only the phase but also polarization. A surface microrelief controlled polarization of an incident beam takes place in polarizing beam splitters [61] and twist-reflectors [62]. Reference [62] reports a solution to the problem of design and fabrication of a twist-reflector that converts the TE_{01} mode of gyrotone radiation with $\lambda = 10\,\mathrm{mm}$ into a plane-polarized wave. The twist-reflector is a curvilinear diffraction grating with rectangular grooves of an equal period less than a wavelength. Similar reflecting diffraction gratings with rectangular grooves have been implemented for IR light with $\lambda = 10.6\,\mu\mathrm{m}$ [155].

The rigorous electromagnetic theory of microrelief diffraction gratings with a period comparable with a wavelength has been developed in [156–159].

Here a problem inverse to that expanded upon in [62] is dealt with. Is it possible solely by modulating the polarization of the incident light wave to form a desired intensity distribution in a certain plane?

The problem is shown to be reducible to the task of the design of the conventional phase DOE. The polarizing DOE can be implemented as a binary phase DOE being a curvilinear diffraction grating or a set of rectangular subgratings with different orientations but with a constant period of less than a wavelength. Since such a grating does not form reflected diffraction orders, one should expect that 80–90% of incident light energy

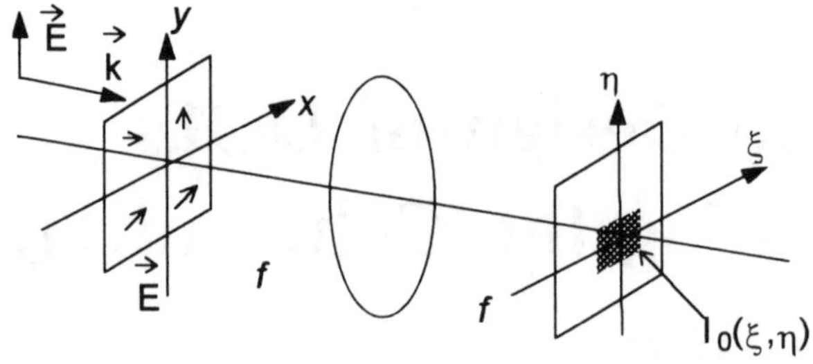

Figure F.1 Optical setup for calculating the polarizing DOE

will concentrate in the domain of the desired image. Since in the conventional binary phase DOE a portion of energy goes to spurious orders, the characteristic energy efficiency is only 40–60%.

An optical setup for the problem under study is shown in Fig. F.1. Let us assume that a plane linearly polarized light wave with the electric vector

$$\vec{\mathbf{E}}_0(x, y) = \vec{\mathbf{e}}_y A_0(x, y)$$

and the amplitude

$$0 \leq A_0(x, y) \leq 1$$

strikes an optical element which, similarly to the thin transparency, instantly changes only the polarization of the wave in every point in the (x, y) plane. Let the electric vector of the light field immediately behind the transparency be given by

$$\vec{\mathbf{E}}(x, y) = E_x(x, y)\vec{\mathbf{e}}_x + E_y(x, y)\vec{\mathbf{e}}_y \tag{F1}$$

where $E_x(x, y)$ and $E_y(x, y)$ are the x- and y-projections of the electric vector, and $\vec{\mathbf{e}}_x$ and $\vec{\mathbf{e}}_y$ are the corresponding axial unit vectors. The lens with focal length f (Fig. F.1) forms in the rear focal plane the desired intensity distribution $I_0(\xi, \eta)$.

Since in a paraxial approximation light complex amplitudes in planes (x, y) and (ξ, η) are connected via a 2D Fourier transform, we can write

$$\begin{aligned} I_0(\xi, \eta) = & \left| \iint_{-\infty}^{\infty} E_x(x, y) \exp\left[\frac{-ik(x\xi + y\eta)}{f}\right] \mathrm{d}x\,\mathrm{d}y \right|^2 \\ & + \left| \iint_{-\infty}^{\infty} E_y(x, y) \exp\left[\frac{-ik(x\xi + y\eta)}{f}\right] \mathrm{d}x\,\mathrm{d}y \right|^2, \quad k = \frac{2\pi}{\lambda} \end{aligned} \tag{F2}$$

Since projections of the electric vector are related through the equation

$$E_x^2(x, y) + E_y^2(x, y) = A_0^2(x, y) \tag{F3}$$

and the linear polarization of light occurs in every point (x, y), we can introduce the following designations

$$E_x(x, y) = A_0(x, y)\cos\varphi(x, y)$$
$$E_y(x, y) = A_0(x, y)\sin\varphi(x, y) \tag{F4}$$

In this case, instead of Eq. (F2) we obtain an equation with respect to the unknown function $\varphi(x, y)$

$$I_0(\xi, \eta) = |\mathfrak{F}\{A_0 \cos\varphi\}|^2 + |\mathfrak{F}\{A_0 \sin\varphi\}|^2 \tag{F5}$$

where $\mathfrak{F}\{\ldots\}$ denotes the Fourier transform. In a number of cases, one can reduce the procedure of solving Eq. (F5) to solving an integral equation for the calculation of the phase DOE

$$I_1(\xi, \eta) = \left| \int_{-\infty}^{\infty}\!\!\int A_0(x, y)\,\mathrm{e}^{i\varphi(x,y)} \exp\left[\frac{-ik(x\xi + y\eta)}{f}\right] \mathrm{d}x\,\mathrm{d}y \right|^2 \tag{F6}$$

where $\varphi(x, y)$ means the light wave phase in the DOE vicinity and it is proportional to the DOE microrelief height.

Equation (F6) can be solved approximately using familiar iterative techniques [2,14,26–28]. From Eqs (F5) and (F6) we can derive relationships between the intensity distributions $I_0(\xi, \eta)$ and $I_1(\xi, \eta)$

$$I_1(\xi, \eta) = I_0(\xi, \eta) + 2\mathrm{Im}\{F_1(\xi, \eta) F_2^*(\xi, \eta)\} \tag{F7}$$

$$I_0 = |F_1(\xi, \eta)|^2 + |F_2(\xi, \eta)|^2 \tag{F8}$$

$$F_1(\xi, \eta) = \mathfrak{F}\{A_0(x, y)\cos\varphi(x, y)\} \tag{F9}$$

$$F_2(\xi, \eta) = \mathfrak{F}\{A_0(x, y)\sin\varphi(x, y)\} \tag{F10}$$

where Im$\{\ldots\}$ is the imaginary part of a number, and * denotes complex conjugation.

Referring to Eq. (F7), it is seen that $I_0(\xi, \eta) = I_1(\xi, \eta)$ if

$$2\,\mathrm{Im}\{F_1 F_2^*\} = 2|F_1||F_2|\sin(\psi_1 - \psi_2) = 0 \tag{F11}$$

where $\psi_{1,2}(\xi, \eta) = \arg F_{1,2}(\xi, \eta)$.

Equation (F11) is met exactly if the function $A_0(x, y)$ is even and $\varphi(x, y)$ is odd, because the Fourier image of these functions is a real function. In this case the phases ψ_1 and ψ_2 in Eq. (F11) can assume only values of $0, +\pi$, and $-\pi$. However, our computer-aided simulation has shown that Eq. (F11) holds approximately even when the $\varphi(x, y)$ function is not necessarily odd.

One can accomplish a turn of the polarization vector through a pregiven angle using, for example, reflecting diffraction gratings with a period less than a wavelength [62,160].

Based on numerically solving the system of equations describing the diffraction of a reflected plane, linearly polarized light wave by diffraction gratings with rectangular grooves and with a period less than a wavelength,

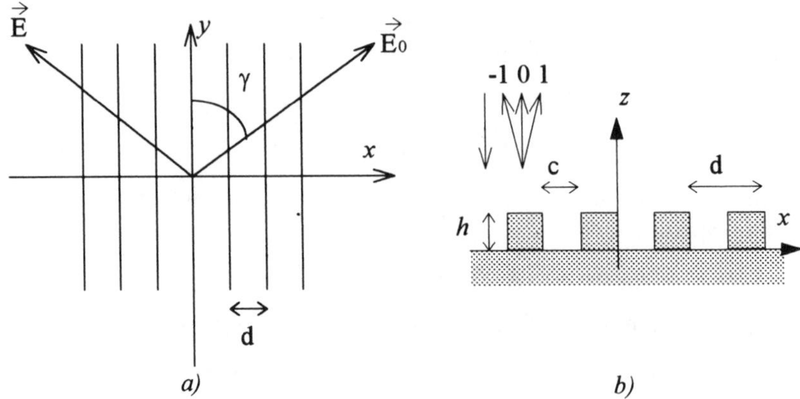

Figure F.2 Binary reflecting grating: turning of polarization vector in (a) top view and (b) side view

it has been shown in [160] that the phase difference between two orthogonal projections of reflected light electric vectors equals π if the grooves' depth equals $\lambda/4$.

This means that the angle between the electric vectors of the incident wave, $\vec{\mathbf{E}}_0$, and the reflected wave, $\vec{\mathbf{E}}$, is 2γ, where γ is the angle between the direction of the grating grooves and the $\vec{\mathbf{E}}_0$ vector (see Fig. F.2a). This can be understood as follows. When the light strikes perpendicularly, the light field within the grooves ($0 \le z \le h$) of a reflective perfectly conducting metal grating (Fig. F.2b) reads [60,155,157]

$$E_y(x,z) = \sum_{n=1}^{\infty} a_n \sin(\mu_n z) \sin\left(\frac{\pi n x}{c}\right) \tag{F12}$$

for the TE polarization (the $\vec{\mathbf{E}}_0$ vector is collinear to the grating grooves and y-axis) and

$$H_y(x,z) = \sum_{n=0}^{\infty} b_n \cos(\mu_n z) \cos\left(\frac{\pi n x}{c}\right) \tag{F13}$$

where

$$\mu_n = \left(\frac{2\pi}{\lambda}\right)^2 - \left(\frac{\pi n}{c}\right)^2 \tag{F14}$$

for the TM polarization.

Referring to Eqs (F12)–(F14) it is seen that if $c \le \lambda/2$, the grating does not produce diffraction orders (except from the exponentially decaying superficial waves), whereas only the field with the electric vector perpendicular to the grating grooves ($0 < x < c$) reflects from the groove bottom

$$H_y(x,z) = b_0 \cos\left(\frac{2\pi z}{\lambda}\right) \tag{F15}$$

Therefore, as a first approximation, we can use the following line of reasoning. The electric vector of an incident light beam can be represented as a sum of two orthogonal components, TE and TM waves, which reflect differently from the diffraction grating: the TE component reflects from the upper faces of grooves, whereas the TM component propagates as far as the groove bottom, undergoes reflection and, in the reverse passage at $z = h$ (Fig. F.2b), interferes with the TE wave. The ray path difference upon the wave interference equals $2h = \lambda/2$, with the phase difference being π.

The optical setup for the work with the polarizing DOE is shown in Fig. F.3. Linearly polarized radiation is entered into the setup via a semi-transparent mirror.

The surface of the polarizing DOE consists of a set of subgratings and is described by the function of microrelief

$$z(x,y) = \frac{\lambda}{8}\sum_{m=1}^{M}\sum_{n=1}^{N} \text{rect}\left[\frac{x - m\Delta_1}{\Delta_1}\right] \text{rect}\left[\frac{y - n\Delta_2}{\Delta_2}\right] \times \text{sgn}\left\{\cos\left[\frac{2\pi(x\cos\bar{\varphi}_{mn} + y\sin\bar{\varphi}_{mn})}{\lambda}\right]\right\} \tag{F16}$$

where

$$\bar{\varphi}_{mn} = \frac{3\pi}{4} + \frac{\varphi_{mn}}{2} \tag{F17}$$

$$\text{rect}\left(\frac{x}{\Delta}\right) = \begin{cases} 1, & |x| \le \dfrac{\Delta}{2} \\ 0, & |x| \ge \dfrac{\Delta}{2} \end{cases} \tag{F18}$$

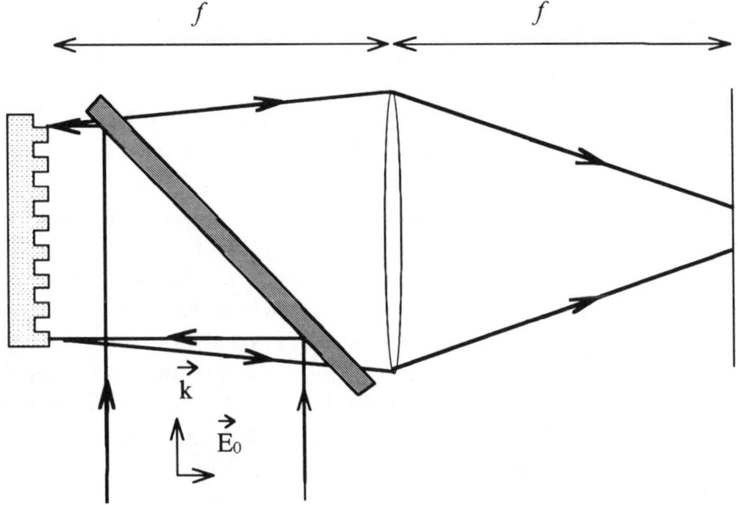

Figure F.3 Optical setup for normal illumination of the polarizing DOE

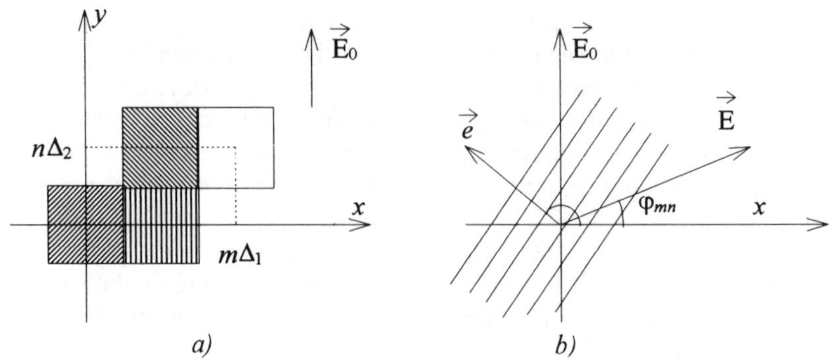

Figure F.4 (a) A fragment of DOE surface composed of subgratings; (b) vector diagram: incident wave vector $\mathbf{E}_0$, reflected wave vector $\mathbf{E}$, and subgrating vector $\vec{\boldsymbol{e}}$

and $\varphi_{mn} = \varphi(x_m, y_n)$ are the phase array pixels resulting from the solution of integral equation (F5).

Figure F.4 depicts the form of the subgratings on the polarizing DOE surface: Δ_1 and Δ_2 are the horizontal and vertical size of the subgrating, $0 < \varphi_{mn} < 2\pi$, $3\pi/4 < \overline{\varphi}_{mn} < 3\pi/4 + \pi$, and $\vec{\mathbf{e}}$ is the unit vector of the subgrating specified by the angle $\overline{\varphi}_{mn}$ (Fig. F.4b).

Note that the transmission dielectric binary grating with a period less than λ and a microrelief height of

$$h = \frac{\lambda}{2(n-1)}$$

(where n is the refractive index of the grating material) also does not produce diffraction orders but just changes the direction of the incident wave polarization vector. This is apparent from [160].

Below we discuss the results of numerically solving Eq. (F6) and subsequently calculating the binary amplitude mask for the fabrication of the polarizing DOE.

The main advantage of the polarizing DOE is as follows: as a binary reflecting or transmitting optical element, it requires for its fabrication the calculation of only one binary amplitude mask, at the same time offering energy efficiency of about 80–90%. This is because the spatial frequency of modulation of the polarizing DOE microrelief is less than a wavelength and the energy does not go into parasite (spurious) diffraction orders, concentrating almost entirely in the area of desired image. Numerical simulation was concerned only with the calculation of the DOE binary microrelief.

A 128×128 network was used for calculating pixels of the phase function $\varphi_{mn} \in [0, 2\pi]$ through the Gerchberg–Saxton method [14]. Twenty-eight iterations gave us an array φ_{mn} shown in Fig. F.5a. If such a mask is utilized

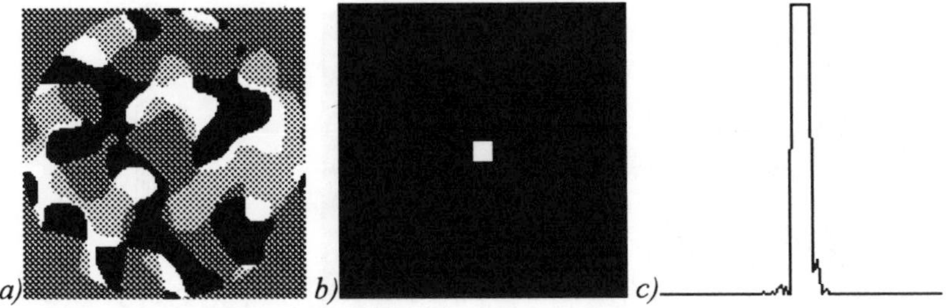

Figure F.5 Phase-modulating DOE; (a) grey-level phase; (b) formed image; (c) image intensity cross-section

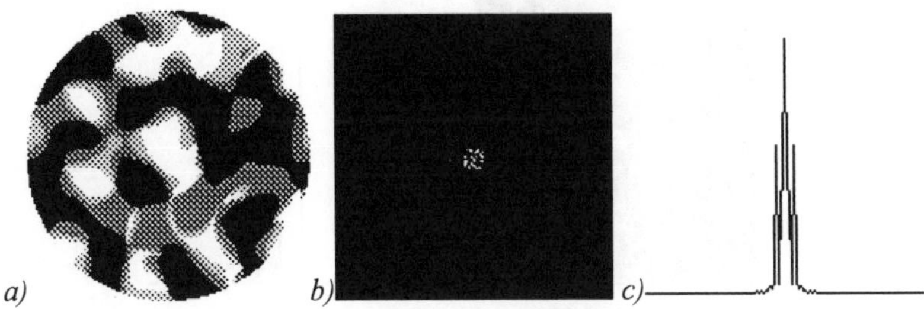

Figure F.6 Stages of calculation of polarizing DOE: (a) mask with transmittance $\cos(\varphi_{mn})$; (b) formed image; (c) its cross-section

in conventional technology to implement a DOE with the grey-level phase function φ_{mn} or transmittance $\exp[i\varphi_{mn}]$, the plane light wave having passed through the DOE will produce in the rear focal plane a light square (Fig. F.5b), with the intensity distribution differing from a pregiven constant value by 0.2% on average. Figure F.5c depicts a horizontal cross-section of the light square intensity. The energy efficiency under which we imply an energy portion coming into a spatial spectrum domain of 8×8 pixels is 82.7%. An initial phase for iterations was chosen to be random.

Next, the solution to Eq. (F6) represented as an array of phases φ_{mn} was substituted in Eq. (F5), with the equation fulfilment to be checked. The procedure is illustrated in Figs F.6 and F.7. Figures F.6a and F.7a show, respectively, arrays $\cos(\varphi_{mn})$ and $\sin(\varphi_{mn})$ (minus one, black colour; plus one, white colour). Figures F.6b and F.7b depict squared modules of the Fourier transform with respect to the $\cos(\varphi_{mn})$ and $\sin(\varphi_{mn})$ functions. In this case the efficiency is 80.3% and 80.1%. The intensity distribution within the squares essentially differs from a constant value (Figs F.6c and F.7c), whereas their sum (Fig. F.8a) is weakly different from the pregiven constant value (Fig. F.8b – intensity cross-section for a total light square), by 2.3% on average.

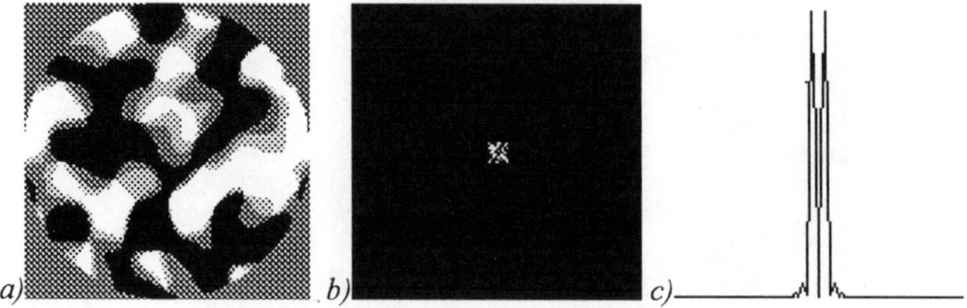

Figure F.7 Stages of calculation of polarizing DOE: (a) mask with transmittance $\sin(\varphi_{mn})$; (b) formed image; (c) its cross-section

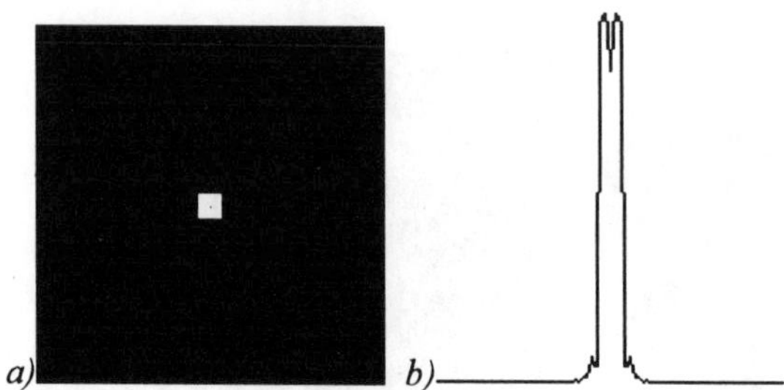

Figure F.8 (a) Aggregate image composed of the pictures shown in Figs F.6b and F.7b; (b) aggregate image's cross-section

Figure F.9a depicts the function describing the polarizing DOE microrelief (binary amplitude mask) derived via Eqs (F16)–(F18) using the phase array φ_{mn} shown in Fig. F.5a, with $\Delta_1 = \Delta_2 = 8$ pixels and 2π in place of λ. A magnified fragment of the amplitude mask (Fig. F.9a) consists of a set of subgratings and is shown in Fig. F.9b.

It is noteworthy that if instead of the initial DOE phase φ_{mn} we take its binary variant $\varphi^{(2)}_{mn} = \{\pi, \text{ if } \pi \leq \varphi_{mn} < 2\pi; \ 0, \text{ if } 0 \leq \varphi_{mn} < \pi\}$ (Fig. F.10a), the resulting light square (Fig. F.10b) will, on average, differ from the required one by 91.1%, with the efficiency decreased to 63.8%.

In the present Appendix we address the problem of designing and fabricating an optical element that would enable us to modulate only the polarization of incident light with the objective of forming a desired image. It has been demonstrated that such an optical element can be calculated using conventional techniques that are commonly used for calculating the DOE's modulating light phase. This element can be fabricated in the form of a binary curvilinear diffraction grating (or a set of subgratings) with a constant period of less than a wavelength.

Figure F.9 (a) Binary mask composed of subgratings of the polarizing DOE forming the square image; (b) its magnified fragment

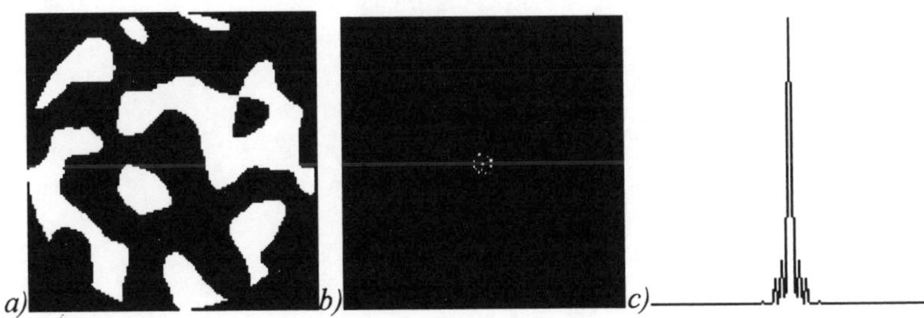

Figure F.10 (a) Binary variant of the initial phase in Fig. F.5a; (b) the light square formed; (c) its cross-section

Appendix G

Iterative Design of Reflection Diffraction Gratings Using the Kirchhoff–Beckmann and Rayleigh Approximations

G.1 A Rayleigh Method for Designing the Grating

Reflection diffraction gratings (DGs) have found use in a variety of applied tasks in optics. Their ability, as beam splitters, not to disturb the beam structure makes them useful in measuring the power of laser beams [161]. They are also employed as diffraction polarizers capable of turning the polarization vector of laser [61] and microwave [62] beams. In optical data processing DGs are used as array illuminators that serve to multiply the laser beam into N beams of equal intensity [43]. There are several approaches, varying in the degree of complexity and accuracy, to the problem of calculating the light field reflected from the diffraction grating (the so-called direct task of diffraction). These are methods of rigorous solution of the electromagnetic Maxwell equations with the corresponding boundary conditions [156], a method for solving coupled-wave equations [159], a Rayleigh method [162], and a method of scalar Kirchhoff diffraction [163].

In the present section, using a planewave representation of diffracted waves (a Rayleigh method), we treat an inverse task of diffraction in which we seek to derive the relief function of the DG from a preset intensity distribution between diffraction orders. The Rayleigh approximation offers the most accurate result under the condition that the maximum height h and the period d of the grating relief are related to the light wavelength λ through the inequalities [164]: $d \leq 15\lambda$, $h \leq 1.5\lambda$.

Following [165], let us briefly consider how the amplitude of light diffracted from the grating is described in the Rayleigh approximation. Figure G.1 depicts an optical configuration for the problem in question. A plane light wave that has the complex amplitude $\psi_i(x, y)$ strikes at an angle of Θ_0 an ideally reflecting DG characterized by the profile changing along the x-axis

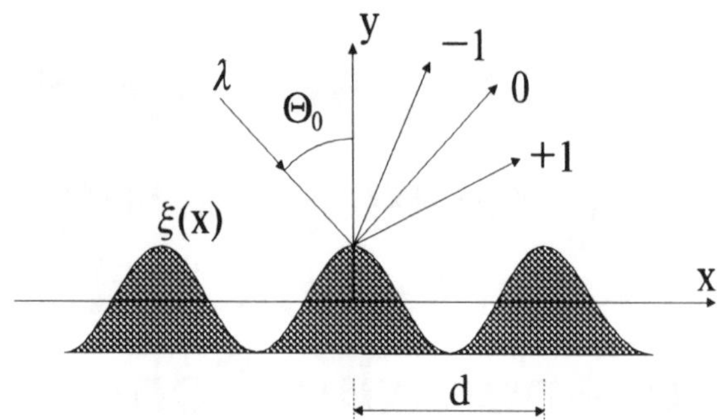

Figure G.1 Optical configuration of the task in question

with a period of d and unchanged along the z-axis (this axis is perpendicular to the plane of Fig. G.1):

$$\psi_i(x, y) = \exp\{ik(x \sin \Theta_0 - y \cos \Theta_0)\} \tag{G1}$$

where $k = 2\pi/\lambda$ is the wavenumber of light.

We confine our consideration to the TE polarization for which the electrical vector of the planewave is directed along the z-axis and the magnetic vector is in the plane of incidence (x, y).

The total light field $\psi(x, y)$ in the half-space over the grating (for $y > \max[\xi(x)]$, where $\xi(x)$ is the grating profile function) satisfies the Helmholtz equation

$$\Delta\psi(x, y) + k^2\psi(x, y) = 0 \tag{G2}$$

with boundary conditions for an ideally reflecting surface

$$\psi(x, y)|_{y=\xi(x)} = 0 \tag{G3}$$

It follows from Eq. (G3) that the light field diffracted by the grating obeys the conditions

$$\psi(x, y) = \psi_d(x, y) + \psi_i(x, y) \tag{G4}$$

$$\psi_d(x, y)|_{y=\xi(x)} = -\exp\{ik(x \sin \Theta_0 - \xi(x) \cos \Theta_0)\} \tag{G5}$$

where $\psi_d(x, y)$ is the diffracted field.

Since the grating function is periodic – $\xi(x + d) = \xi(x)$ – and the function in the right-hand side of Eq. (G5) is quasi-periodic, the following function should be periodic in x

$$V(x, y) = \psi_d(x, y) \exp\{-ikx \sin \Theta_0\} \tag{G6}$$

Next, we expand the function in Eq. (G6) into a Fourier series

$$V(x, y) = \sum_{n=-\infty}^{\infty} V_n(y) \exp\left\{2\pi i \frac{nx}{d}\right\} \tag{G7}$$

From Eqs (G6) and (G7) we derive the planewave representation for the amplitude of the diffracted field

$$\psi_d(x, y) = \sum_{n=-\infty}^{\infty} V_n(y) \exp\{ik\alpha_n x\} \tag{G8}$$

where

$$\alpha_n = \sin\Theta_0 + n\frac{\lambda}{d}$$

Substituting Eq. (G8) into the Helmholtz equation, we find that

$$\sum_{n=-\infty}^{\infty}\left[\frac{\mathrm{d}^2 V_n(y)}{\mathrm{d}y^2} + k^2\beta_n^2 V_n(y)\right]\exp\left\{2\pi i\frac{nx}{d}\right\} = 0 \tag{G9}$$

from which there follow equations for the functions $V_n(y)$

$$\frac{\mathrm{d}^2 V_n(y)}{\mathrm{d}y^2} + k^2\beta_n^2 V_n(y) = 0 \tag{G10}$$

where $\beta_n^2 = 1 - \alpha_n^2$.

The solution to Eq. (G10) takes the form

$$V_n(y) = B_n \exp\{ik\beta_n y\} \tag{G11}$$

Equation (G8), in view of Eq. (G11), leads to the following form of the decomposition of the diffracted field in terms of planewaves at $y > \max[\xi(x)]$

$$\psi_d(x, y) = \sum_{n=-\infty}^{\infty} B_n\psi_n(x, y) \tag{G12}$$

$$\psi_n(x, y) = \exp\{ik(\alpha_n x + \beta_n y)\} \tag{G13}$$

Note that the sum in Eq. (G12) contains both homogeneous planewaves (at $\alpha_n^2 < 1$) and inhomogeneous waves exponentially damped along the y-axis under the conditions that $\alpha_n^2 > 1$.

The Rayleigh hypothesis assumes that the expansion in Eq. (G12) is true not only for $y > \max[\xi(x)]$, but also for $y \geq \xi(x)$. Equations (G5) and (G12) give an expression for deriving the coefficients B_n

$$\sum_{n=-\infty}^{\infty} B_n\psi_n[x, \xi(x)] = -\psi_i[x, \xi(x)] \tag{G14}$$

If we restrict the number of diffracted orders, $-N < n < N$, and choose a set of points x_m, Eq. (G14) leads to a linear set of algebraic equations in B_n

$$\sum_{n=-N}^{N} B_n\psi_{nm} = -\psi_{im} \tag{G15}$$

where $\psi_{nm} = \psi_n[x_m, \xi(x_m)]$, $\psi_{im} = \psi_i[x_m, \xi(x_m)]$.

Below we deal with an inverse task and propose an iterative algorithm for calculating the functions $\xi(x)$ using preset modules of the coefficients B_n. Squared modules of the coefficients $|B_n| = I_{n0}$ represent the intensities of diffracted orders to be generated. We propose that this task should be solved as follows. From Eq. (G14) the function $\psi_d(x, y)$ is seen to be phase-only at $y = \xi(x)$

$$|\psi_d[x, \xi(x)]| = |\psi_i[x, \xi(x)]| = 1$$

Making use of the indeterminacy of arguments of the complex coefficients B_n in the sum of Eq. (G12), we can iteratively calculate the diffracted field that would have a unit module in a rectangular domain

$$-\frac{d}{2} < x < \frac{d}{2}, -\frac{\lambda}{2} < y < \frac{\lambda}{2}$$

The calculational algorithm is similar to the familiar Gerchberg–Saxton algorithm [14] and involves the following steps.

1. Assume that in the pth iteration step the estimate of the function of the diffracted field is given by

$$\psi_d^{(p)}(x, y) = \exp\{iQ_p(x, y)\} \tag{G16}$$

2. The function in Eq. (G16) can be expanded into a series in orthogonal functions of Eq. (G13), the series coefficients being found from the relation

$$B_n^{(p)} = \frac{1}{\lambda d} \int_0^d \int_{-\lambda/2}^{\lambda/2} \exp\{iQ_p(x, y) - ik(\alpha_n x + \beta_n y)\}\, \mathrm{d}x\, \mathrm{d}y \tag{G17}$$

3. The calculated coefficients B_n are modified in view of preset values of intensity in orders

$$\bar{B}_n^{(p)} = \sqrt{(I_{n0})}\, \exp\{i \arg(B_n^{(p)})\} \tag{G18}$$

4. The next estimate of phase of the diffracted field function is found using the relation

$$Q_{p+1}(x, y) = \arg\left\{ \sum_{n=-\infty}^{\infty} \bar{B}_n^{(p)} \exp[ik(\alpha_n x + \beta_n y)] \right\} \tag{G19}$$

5. Next is the passage to step 1, and so on.

Assume that after a certain number of steps the r.m.s. deviation has achieved a preset value δ_0:

$$\delta_p = \frac{1}{\lambda d}\sqrt{\left\{\int_0^d \int_{-\lambda/2}^{\lambda/2} [|\psi_d^{(p)}(x,y)| - 1]^2 \mathrm{d}x\,\mathrm{d}y\right\}} \leq \delta_0 \tag{G20}$$

Then, based on Eq. (G14), we can write an equation serving to find the relief function $\xi(x)$

$$Q_p[x, \xi(x)] = \pi + k\alpha_0 x - k\beta_0 \xi(x) \tag{G21}$$

where $Q_p[x, \xi(x)]$ is the phase of the diffracted field $\psi_d(x, y)$ that has been found in the pth iteration.

Figure G.2 shows that the solution to Eq. (G21) for $\Theta_0 = 0$ ($\alpha_0 = 0$, $\beta_0 = 1$) and for any point x_0 can be found from the intersection of the straight line

$$f_1(y) = -\frac{2\pi}{\lambda}y$$

and the curve

$$f_2(y) = Q_p(x_0, y) - \pi \tag{G22}$$

G.2 Kirchhoff–Beckmann Approximation for Designing the Relief Surface

Diffractive optical elements capable of generating a pattern consisting of light spots (dots) of uniform intensity are necessary for the devices for parallel data processing [43,166]. Such DOEs are called array illuminators (AIs). As

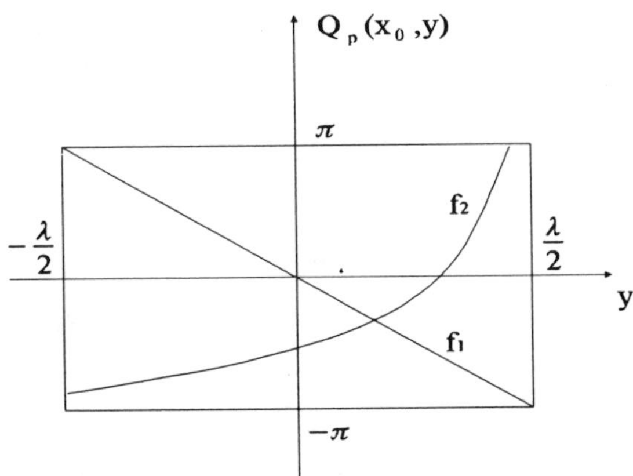

Figure G.2 Graphic solution to Eq. (G22)

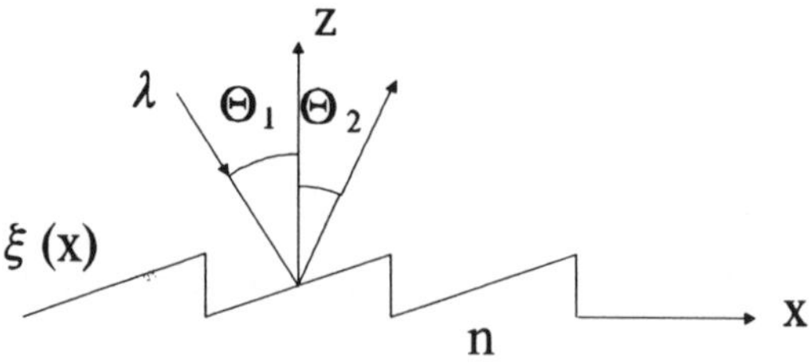

Figure G.3 Optical diagram

AIs, one usually uses binary phase diffraction gratings, such as Dammann gratings [40], that have high energy efficiency: about 70–80% of the light energy striking the element contributes to the generation of the required number of diffraction orders. The number of orders may be as many as 128×128, whereas the deviation from uniform intensity in orders does not exceed 10% [43]. The design of binary phase AIs proceeds, as a rule, either using Fourier optics methods [43] or using rigorous electromagnetic theory [156]. The former implies designing the AI in the approximation of a thin phase transparency that makes the incident light field undergo some phase shift in each point of the transparency. The latter involves designing the AI as a reflecting surface of infinite conductivity and with a relief in the form of rectangular grooves. Note, however, that the authors were unable to find anywhere in the literature the formulation of the inverse task within the framework of the rigorous electromagnetic approach. In this Appendix we propose an iterative gradient method for designing reflecting AIs of finite conductivity using Kirchhoff's theory. It has been shown numerically that blazed AIs (with a sawtooth relief) are a convenient interim tool for obtaining binary AIs.

Reference [167] reports obtaining a Kirchhoff's diffraction approximation of the expression for the angular distribution $\rho(\Theta_2)$ of complex amplitude that describes the far field diffraction caused by the reflection of the planewave from a surface with the smooth relief that has a height less than a wavelength and is described by the function of type $\xi(x)$ (see Fig. G.3)

$$\rho(\Theta_2) = \frac{1}{4L\cos\Theta_1}\int_{-L}^{L}[a(x)\,\xi'(x) - b(x)]\exp\{i[s_x x + s_z\xi(x)]\}\,dx \tag{G23}$$

where

$$a(x) = (1 - r_F)\sin\Theta_1 + (1 + r_F)\sin\Theta_2$$

$$b(x) = (1 + r_F)\cos\Theta_2 - (1 - r_F)\cos\Theta_1$$

$$s_x = k(\sin\Theta_1 - \sin\Theta_2)$$

$$s_z = -k(\cos\Theta_1 + \cos\Theta_2)$$

$$k = \frac{2\pi}{\lambda} \text{ is the wavenumber of light}$$

λ is the wavelength, $[-L, L]$ is the illuminated portion of the surface, Θ_1 and Θ_2 are the angles of incidence and reflection, and r_F is Fresnel's coefficient of reflection, which, for example, for a TE wave (the electric vector is perpendicular to the plane of incidence, see Fig. G.3) is given by [3]

$$r_{\mathrm{FTE}} = \frac{\cos\theta - \sqrt{(\tilde{n}^2 - \sin^2\theta)}}{\cos\theta + \sqrt{(\tilde{n}^2 - \sin^2\theta)}} \tag{G24}$$

where $\theta = \Theta_1 - \tan^{-1}\xi'(x)$ is the local angle of incidence, $\xi'(x)$ is the derivative of the surface relief function, $\tilde{n} = n + i\tilde{k}$ is the complex index, n is the refractive index of the surface, and $\tilde{k}$ is the absorption factor of the surface.

We state the problem of deriving the function $\xi(x)$ from the preset angular intensity distribution $i(\Theta_2) = |\rho_0(\Theta_2)|^2$ as a variation task on finding the minimum of the residual functional

$$I = \int_{-\pi/2}^{\pi/2} [|\rho_0(\Theta_2)| - |\rho(\Theta_2, \xi, \xi')|]^2 \mathrm{d}\Theta_2 \tag{G25}$$

where $|\rho_0|$ and $|\rho|$ are the preset and the estimated module of the complex angular amplitude. We shall find the extremum of the criterion in Eq. (G25) using the Rayleigh–Ritz method [110]. For this purpose, the desired function $\xi(x)$ of relief may be approximately represented as a linear combination of some familiar basis functions

$$\xi(x) \approx \sum_{n=1}^{N} c_n \psi_n(x) \tag{G26}$$

Next, the coefficients of the sum in Eq. (G26) are considered as parameters, by the fitting of which the functional in Eq. (G25) is minimized. The coefficients c_n are found using the familiar iterative procedure

$$c_n^{(k+1)} = c_n^{(k)} - \tau_n \frac{\partial I}{\partial c_n^{(k)}}, \quad n = \overline{1, N} \tag{G27}$$

where $c_n^{(k)}$ and $c_n^{(k+1)}$ are the coefficients of the sum in Eq. (G26) derived in the kth and $(k+1)$th steps, and τ_n is the step of the nth equation that is sought via a conventional linear procedure [74] from the condition of the local minimum of gradient of the function in Eq. (G25). The derivatives entering into Eq. (G27) take the form

$$\frac{\partial I}{\partial c_n} = \int_{-\pi/2}^{\pi/2} \left(1 - \frac{|\rho_0|}{|\rho|}\right)\left(\rho^* \frac{\partial \rho}{\partial c_n} + \rho \frac{\partial \rho^*}{\partial c_n}\right) \mathrm{d}\Theta_2 \tag{G28}$$

$$\begin{aligned} \frac{\partial \rho}{\partial c_n} = \frac{1}{4L\cos\Theta_1} \int_{-L}^{L} & \left(\xi' \frac{\partial a}{\partial c_n} + a \frac{\partial \xi'}{\partial c_n} - \frac{\partial b}{\partial c_n} + (ia\xi' - ib) s_z \frac{\partial \xi}{\partial c_n}\right) \\ & \times \exp\{i[s_x x + s_z \xi(x)]\}\, \mathrm{d}x \end{aligned} \tag{G29}$$

$$\frac{\partial a}{\partial c_n} = (\sin \Theta_2 - \sin \Theta_1) \frac{\partial r_F}{\partial c_n} \tag{G30}$$

$$\frac{\partial b}{\partial c_n} = (\cos \Theta_2 + \cos \Theta_1) \frac{\partial r_F}{\partial c_n} \tag{G31}$$

$$\frac{\partial \xi}{\partial c_n} = \psi_n \tag{G32}$$

$$\frac{\partial \xi'}{\partial c_n} = \frac{\mathrm{d}\psi_n}{\mathrm{d}x} \tag{G33}$$

$$\frac{\partial r_F}{\partial c_n} = \frac{2r_F \sin \theta}{\cos \theta + \sqrt{(\tilde{n} - \sin^2 \theta)}} \frac{\partial \theta}{\partial c_n} \tag{G34}$$

$$\frac{\partial \theta}{\partial c_n} = [1 + (\xi')^2] \frac{\partial \xi'}{\partial c_n} \tag{G35}$$

We consider a special case of the above method applied to a sawtooth reflecting surface. Assume that the illuminated portion $[-L, L]$ of the surface is broken down into N equal segments 2Δ in length on which the sawtooth relief does not change the angle of inclination ω_n of the generator. Then the function $\xi(x)$ can be written as

$$\xi(x) = \sum_{n=1}^{N} \mathrm{rect}\left(\frac{x - 2n\Delta}{\Delta}\right) [\alpha_n(x - 2n\,\Delta) + \varphi_n] \tag{G36}$$

where $\alpha_n = \tan \omega_n$, φ_n is the magnitude of displacement along the z-axis from some plane (see Fig. G.4). Substituting the function of sawtooth relief of Eq. (G36) into Eq. (G23) yields a specific form of the angular distribution of amplitude

$$\rho(\Theta_2) = \sum_{n=1}^{N} Q_n(\Theta_2) \exp\{-i\gamma_n g(\Theta_2)\} \tag{G37}$$

where

$$\begin{aligned} Q_n(\Theta_2) = {} & \frac{\Delta \sqrt{(1 + \alpha_n^2)}}{2L \cos \Theta_1} [(1 - r_n) \cos(\Theta_1 - \omega_n) - (1 + r_n) \cos(\Theta_2 + \omega_n)] \\ & \times \mathrm{sinc}\left\{ 2k\Delta \sqrt{(1 + \alpha_n^2)} \cos\left(\frac{\Theta_1 + \Theta_2}{2}\right) \sin\left(\frac{\Theta_1 - \Theta_2}{2} + \omega_n\right)\right\} \\ & \times \exp\{2ink\Delta(\sin \Theta_1 - \sin \Theta_2)\} \end{aligned} \tag{G38}$$

$$\gamma_n = \frac{\varphi_n}{\Delta}, \quad g(\Theta_2) = k\Delta(\cos \Theta_1 + \cos \Theta_2)$$

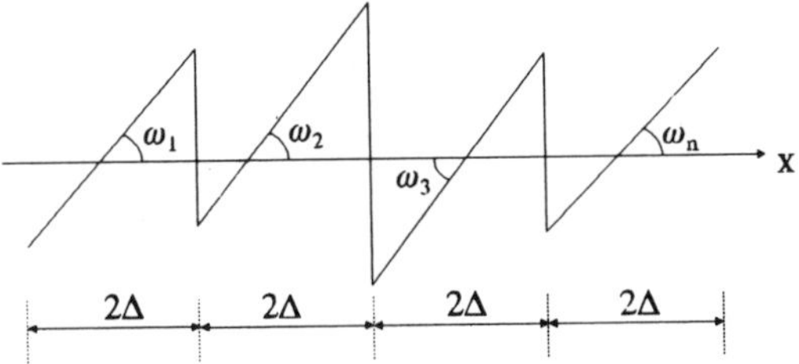

Figure G.4 A sawtooth surface

r_n is Fresnel's coefficient of reflection r_F in the nth segment of the surface.

Designating $g(\Theta_2) = x$, instead of Eq. (G37) we obtain

$$\rho(x) = \sum_{n=1}^{N} Q_n(x) \exp\{-i\gamma_n x\} \tag{G39}$$

Equation (G39) shows that the function $\rho(\Theta_2)$ is a result of superposition of contributions from each segment of the breakdown. Each individual contribution is proportional to the sinc-function which attains its maximum value at the diffraction angle of $\Theta_{2n} = 2\omega_n + \Theta_1$, the effective angular width of the sinc-function being proportional to the value of

$$\left[\frac{2k\Delta}{\sin\omega_n}\sin(\omega_n + \Theta_1)\right]^{-1}$$

If one chooses the angles of inclination ω_n of the relief generator in such a manner that all diffraction orders (the terms in the sum of Eq. (G39)) are spatially separated in the far field, the synthesis of the surface of an AI appears to be trivial. The surface will take the shape shown in Fig. G.5a. The entire surface in this example is broken down into three equal segments, in each of which the angle of inclination of the sawtooth relief is constant and the relief height is the same and equals $\lambda/2$. The calculation of diffraction of the plane TE wave by such a surface ($\Theta_1 = 0$) has shown that there occur three diffraction orders of equal intensity (Fig. G.5b). With no special conditions imposed on the angles ω_n, the problem of the design of blazed AIs can be formulated in the general case. Assume that the parameters Δ and α_n are fixed and the only free parameter is γ_n. Let us once again consider the variation task on the determination of maximum of the function in Eq. (G25) which is to depend on the parameters γ_n

$$I(\gamma_n) = \int_{-\pi/2}^{\pi/2} [|\rho_0(\Theta_2)| - |\rho(\Theta_2, \gamma_n)|]^2 \mathrm{d}\Theta_2 \tag{G40}$$

In this case, instead of the gradient procedure (see Eq. (G27)) we have

$$\gamma_n^{(k+1)} = \gamma_n^{(k)} - \tau_n \frac{\partial I}{\partial \gamma_n^{(k)}}, \quad n = \overline{1, N} \tag{G41}$$

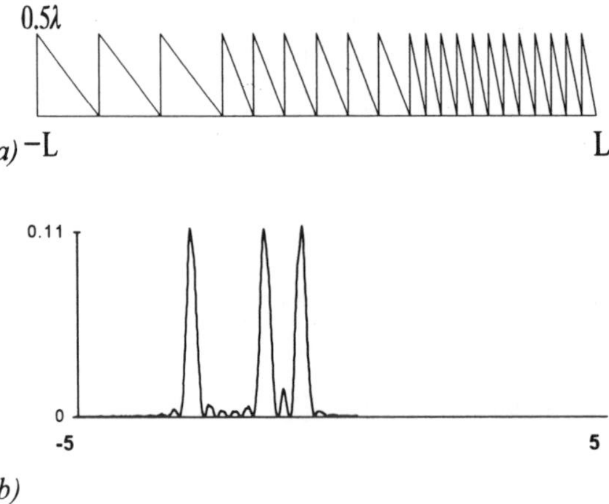

Figure G.5 A sawtooth relief (a) and diffraction orders (b)

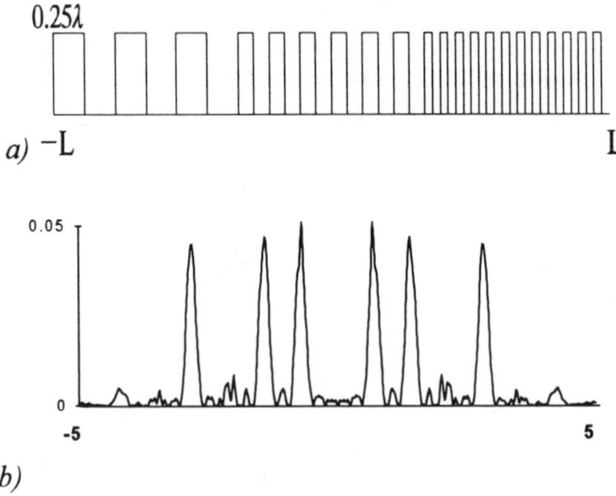

Figure G.6 Binary relief (a) and diffraction orders (b)

where

$$\frac{\partial I}{\partial \gamma_n} = \int_{-\pi/2}^{\pi/2} \left(1 - \frac{|\rho_0|}{|\rho|}\right)\left(\rho^* \frac{\partial \rho}{\partial \gamma_n} + \rho \frac{\partial \rho^*}{\partial \gamma_n}\right) \mathrm{d}\Theta_2 \tag{G42}$$

$$\frac{\partial \rho}{\partial \gamma_n} = -i \sum_{n=1}^{N} Q_n(\Theta_2) g(\Theta_2) \exp\{-i\gamma_n g(\Theta_2)\} \tag{G43}$$

Optimal values for the steps τ_n in the Eqs (G41) are found using a linear search of the maximum value of derivatives $\partial I/\partial \gamma_n$ in each iteration step (see Eq. (5.18)).

The following example shows that AIs with sawtooth relief are convenient to use. Figure G.6a depicts the binary profile of the reflection AI that has a relief height of $\lambda/4$ and has been produced via a binarization of the sawtooth relief shown in Fig. G.5a. Figure G.6b illustrates the result of calculation of the plane TE wave diffraction by the surface having the relief shown in Fig. G.6a. Ratios of the surface scattered intensity to the total intensity of the incident light are plotted as ordinate, and angles of scattering, measured in degrees, as abscissa. Instead of three orders of equal intensity (Fig. G.5b) we find here six orders, also of equal intensity (Fig. G.6b). The simulation parameters are: $L = 0.25$ mm, $\lambda = 0.63\ \mu$m, $\tilde{n} = n + i\tilde{k}$, $n = \infty$, $\tilde{k} = 0$ is the situation of infinite conductivity, $\Theta_1 = 0$, and $M = 512$ is the general number of pixels. The minimum period of the relief modulation is about 10λ.

References

[1] Lesem L. B., Hirsh P. M., Jordan J. A. The kinoform: new wavefront reconstruction device. IBM J. Res. Dev. 1969, v. 13, no. 2, pp. 150–155.

[2] Hirsh P. M., Jordan J. A., Lesem L. B. Method of marking an object dependent diffuser. US Patent 3,619,022, November, 1971.

[3] Born M., Wolf E. *Principles of optics*, Pergamon Press, Oxford, 1970.

[4] Optical Engineering, 1993, v. 32, no. 10.

[5] *Inverse scattering problems in optics*. Ed. Baltes H. P., Springer Verlag, Berlin, 1980.

[6] *Image recovery: theory and application*. Ed. Stark H., Academic Press, New York, 1987.

[7] Bates R., McDonnel M. *Image restoration and reconstruction*, Pergamon Press, Oxford, 1986.

[8] Bringdhal O., Wyrowski F. *Digital holography – computer-generated holograms*. Progress in Optics, Ed. Wolf E., v. 28, pp. 1–86, North-Holland, Amsterdam, 1990.

[9] Lohmann A. W., Paris B. P. Binary Fraunhofer holograms generated by computer. Appl. Opt., 1967, v. 6, no. 10, p. 1739.

[10] Brown B. R., Lohmann A. W. Complex spatial filtering with binary masks. Appl. Opt., 1966, v. 5, no. 6, pp. 967–969.

[11] Lee W. H. *Computer-generated holograms: techniques and applications*. Progress in Optics, Ed. Wolf E., North-Holland, Amsterdam, 1978.

[12] Chu D. C., Goodman J. W. Spectrum shaping with parity sequences. Appl. Opt., 1972, v. 11, no. 8, pp. 1716–1724.

[13] Kirk J. P., Jones A. L. Phase only complex-valued spatial filter. J. Opt. Soc. Am., 1971, v. 61, no. 8, pp. 1023–1028.

[14] Gerchberg R. W., Saxton W. O. A practical algorithm for the determination of the phase from image and diffraction plane pictures. Optik, 1972, v. 35, no. 2, pp. 237–246.

[15] Lin B., Gallagher N. C. Convergence of a spectrum shaping algorithm. Appl. Opt., 1974, v. 13, no. 11, pp. 2470–2471.

[16] GALLAGHER N. C., LIN B. Method for computing kinoforms that reduces image reconstruction error. Appl. Opt., 1973, v. 12, no. 10, pp. 2328–2335.

[17] CHU D. C., FIENUP J. R. Recent approaches to computer-generated holograms. Opt. Eng., 1974, v. 13, no. 3, pp. 189–195.

[18] FIENUP J. R. Reconstruction of an object from the modulus of its Fourier transform. Opt. Lett., 1978, v. 3, no. 1, pp. 27–29.

[19] FIENUP J. R. Iterative method applied to image reconstruction and to computer generated holograms. Opt. Eng., 1980, v. 19, no. 3, pp. 297–305.

[20] FIENUP J. R. Phase retrieval algorithm: a comparison. Appl. Opt. 1982, v. 21, no. 15, pp. 2758–2769.

[21] WEISSBACH S., WYROWSKI F., BRYNGDAHL O. Digital phase holograms: coding and quantization with an error diffusion concept. Opt. Commun., 1989, v. 72, no. 1, 2, pp. 37–41.

[22] WEISSBACH S., WYROWSKI F., BRYNGDAHL O. Error-diffusion algorithm in phase synthesis and retrieval techniques. Opt. Lett., 1992, v. 17, no. 4, pp. 235–237.

[23] WYROWSKI F., BRYNGDAHL O. Iterative Fourier-transform algorithm applied to computer holography. J. Opt. Soc. Am. A, 1988, v. 5, no. 7, pp. 1058–1065.

[24] BROJA M., WYROWSKI F., BRYNGDAHL O. Digital halftonic by iterative procedure. Opt. Commun., 1989, v. 60, no. 3, 4, p. 205.

[25] WYROWSKI F. Diffractive optical elements: iterative calculation of quantized, blazed phase structures. J. Opt. Soc. Am. A, 1990, v. 7, no. 6, pp. 961–963.

[26] KOTLYAR V. V., NIKOLSKI I. V., SOIFER V. A. Adaptive iterative algorithm for focusators synthesis. Optik, 1991, v. 88, no. 1, pp. 17–19.

[27] KOTLYAR V. V., NIKOLSKI I. V. Iterative computing of transmittance of optical elements focusing at a predetermined area. Opt. Las. Eng., 1991, v. 15, no. 5, pp. 323–330.

[28] KHONINA S. N., KOTLYAR V. V., SOIFER V. A. Fast Hankel transform for focusators synthesis. Optik, 1991, v. 88, no. 4, pp. 182–184.

[29] KHONINA S. N., KOTLYAR V. V., SOIFER V. A. Calculation of the focusator into a longitudinal line-segment and study of a focal area. J. Mod. Opt., 1993, v. 40, no. 5, pp. 761–769.

[30] KHONINA S. N., KOTLYAR V. V., SOIFER V. A. Algorithm for the generation of non-diffracting Bessel modes. J. Mod. Opt., 1995, v. 42, no. 6, pp. 1231–1239.

[31] KOTLYAR V. V., NIKOLSKI I. V., SOIFER V. A. An algorithm for calculation of formers of Gaussian modes. Optik, 1994, v. 98, no. 1, pp. 26–30.

[32] KAZANSKIY N. L., KOTLYAR V. V., SOIFER V. A. Computer-aided design of diffractive optical elements. Opt. Eng., 1994, v. 33, no. 10, pp. 3156–3166.

[33] GOLUB M. A., KARPEEV S. V., PROKHOROV A. M., SISAKYAN I. N., SOIFER V. A. Focusing of light into a specified volume by computer synthesized holograms. Sov. Techn. Phys. Lett., 1981, v. 7, no. 5, pp. 264–266.

[34] DANILOV V. A., POPOV V. V., PROKHOROV A. M., SAGATELYAN D. M., SISAKYAN I. N., SOIFER V. A. Synthesis of optical elements generating an arbitrary form focal line. Sov. Techn. Phys. Lett., 1982, v. 8, no. 6, pp. 351–353.

[35] SISAKYAN I. N., SOIFER V. A. Computer optics: achievements and problems.

Comput. Opt., 1989, v. 1, no. 1, pp. 3–12.

[36] Golub M. A., Degtyareva V. P., Klimov A. N., Popov V. V., Prokhorov A. M., Sisakyan E. V., Sisakyan I. N., Soifer V. A. Computer synthesis of focusing elements for a CO_2-laser. Sov. Techn. Phys. Lett., 1982, v. 8, no. 4, pp. 195–196.

[37] Golub M. A., Sisakyan I. N., Soifer V. A. Infrared radiation focusators. Opt. Las. Eng., 1991, v. 15, no. 5, pp. 297–309.

[38] Sisakyan I. N., Soifer V. A. Infrared focusators, new optical elements. Infrared Phys., 1991, v. 32, pp. 435–438.

[39] Akahori H. Spectrum leveling by an iterative algorithm with a dummy area for synthesizing the kinoform. Appl. Opt., 1986, v. 25, no. 5, pp. 802–811.

[40] Dammann H., Gortler K. High-efficiency in-line multiple imaging by means of multiple phase holograms. Opt. Commun., 1971, v. 3, no. 5, pp. 312–315.

[41] Bobrov S. T., Turkevich Yu G. Multiorder diffraction gratings with asymmetric profile of a period. Comput. Opt., Moscow, 1989, no. 4, pp. 38–45.

[42] Bereznyi A. E., Komarov S. V., Prokhorov A. M., Sisakyan I. N., Soifer V. A. Phase diffraction gratings with prescribed parameters: inverse problem in optics. Sov. Phys. Dokl., 1986, v. 31, no. 3, pp. 260–263.

[43] Vasara A., Taghizadeh M. R., Turunen J., Westerholm J., Noponen E., Ichikawa H., Miller J. R., Jaakkola T., Kuisma S. Binary surface-relief gratings for array illumination in digital optics. Appl. Opt., 1992, v. 31, no. 17, pp. 3320–3336.

[44] Roberts N. C., Kirk A. G., Hall T. J. Binary phase gratings for hexagonal array generation. Opt. Commun., 1992, v. 94, no. 6, pp. 501–505.

[45] Feldman M. R., Gest C. C. Iterative encoding of high-efficiency holograms for generation of spot arrays. Opt. Lett., 1989, v. 14, no. 8, pp. 479–481.

[46] Wyrowski F. Modulation schemes of phase gratings. Opt. Eng., 1992, v. 31, no. 2, pp. 251–257.

[47] Golub M. A., Doskolovich L. L., Kazanskiy N. L., Kharitonov S. I., Soifer V. A. Computer-generated diffractive multi-focal lens. J. Mod. Opt., 1992, v. 39, no. 6, pp. 1245–1251.

[48] Khonina S. N., Kotlyar V. V., Soifer V. A. Diffraction computation of a focusator into a longitudinal segment and multifocal lens. Proc. SPIE, 1993, v. 1780, pp. 263–272.

[49] McLeod J. H. The axicon: a new type optical element. J. Opt. Soc. Am. 1954, v. 44, no. 8, pp. 592–597.

[50] Khonina S. N., Kotlyar V. V., Shinkaryev M. V., Soifer V. A., Uspleniev G. V. Trochoson. Opt. Commun., 1992, v. 91, no. 3, 4, pp. 158–162.

[51] Bobrov S. T., Greysuch G. I., Turkevich Yu. G. *Optics of diffractive elements and systems*, Mashinostroyeniye, Leningrad, 1986.

[52] Golub M. A., Zhivopistsev E. S., Karpeev S. V., Prokhorov A. M., Sisakyan I. N., Soifer V. A. Obtaining aspherical wavefronts with computer holograms. Sov. Phys. Dokl., 1980, v. 25, no. 8, pp. 627–629.

[53] Golub M. A., Kazanskiy N. L., Sisakyan I. N., Soifer V. A. Etalon synthesis for controlling the off-axis segments of the aspherical surface. Opt. Spectrosc., 1990, v. 68, no. 2, pp. 461–466.

[54] DURNIN J., MICELI J. J., EBERLY J. H. Diffraction-free beams. Phys. Rev. Lett., 1987, v. 58, no. 15, pp. 1499–1501.
[55] VASARA A., TURUNEN J., FRIBERG A. T. Realization of general non-diffracting beams with computer-generated holograms. J. Opt. Soc. Am. A, 1989, v. 6, pp. 1748–1754.
[56] SOIFER V. A., GOLUB M. A. *Laser beam mode selection by computer generated holograms*, CRC Press, Boca Raton, FL, 1994.
[57] LESEBERG D. Sizable Fresnel holograms generated by computer. J. Opt. Soc. Am. A, 1989, v. 6, no. 2, pp. 229–233.
[58] KOTLYAR V. V., NIKOLSKI I. V., SOIFER V. A. Fast calculation of large-dimensional focusators. Pure Appl. Opt., 1994, v. 3, no. 1, pp. 37–44.
[59] DOSKOLOVICH L. L., SOIFER V. A., SHINKARYEV M. S. A method for stochastically synthesizing binary diffraction gratings. Avtometriya, 1992, v. 3, pp. 104–108 (in Russian).
[60] KOK Y., GALLAGHER N. C. Relative phase of electromagnetic wave diffracted by a perfectly conducting rectangular-grooved grating. J. Opt. Soc. Am. A, 1988, v. 5, no. 1, pp. 65–73.
[61] HAIDNER H., KIPFER P., SHERIDAN J. T., SCHWIDER J., STREIBL N., LINDORF J., COLLISCHON M., LONG A., HUTFLESS J. Polarizing reflection grating beamsplitter for 10.6 μm wavelength. Opt. Eng., 1993, v. 32, no. 8, pp. 1860–1865.
[62] SWEENEY D. M., GALLAGHER N. C. Multielement microwave computer generated holograms. Proc. SPIE, 1988, v. 884, pp. 114–119.
[63] BEREZNYI A. E., PROKHOROV A. M., SISAKYAN I. N., SOIFER V. A. Bessel optics. Sov. Phys. Dokl., 1984, v. 29, no. 2, pp. 115–117.
[64] KHONINA S. N., KOTLYAR V. V., SHINKARYEV M. V., SOIFER V. A., USPLENIEV G. V. The phase rotor filter. J. Mod. Opt., 1992, v. 39, no. 5, pp. 1147–1154.
[65] GREYSUCH G. I., UFIMENKO I. M., STEPANOV S. A. *Optics of gradient and diffractive elements*, Radio i svyaz, Moscow, 1990 (in Russian).
[66] KORONKEVICH V. P., MICHATZOV I. A., CHURIN Y. G., YURLOV Y. I. Diffractive elements for pointing laser beams. Avtometriya, 1994, v. 3, pp. 57–68 (in Russian).
[67] KORONKEVICH V. P., NAGORNI V. N., PALCHIKOVA I. V., POLESCHUK A. G. Bifocus microscope. Optik, 1988, v. 78, no. 2, pp. 64–68.
[68] LESEBERG D. Computer generated holograms: cylindrical, conical and helical waves. Appl. Opt., 1987, v. 26, no. 20, p. 4385.
[69] TREMBLAY R., D'ASTONS Y., ROY G., BLANSHARD M. Laser plasmas optically pumped by focusing with axicon a CO_2–TEA laser beam in a high-pressure gas. Opt. Commun., 1979, v. 28, no. 2, p. 193.
[70] GOODMAN J. W. *Introduction to Fourier optics*, McGraw-Hill, New York, 1968.
[71] SIVOKON V. P. Formation of light beams with pregiven structure for the tasks of laser technology. Master's Thesis, Moscow State University Publishers, 1986 (in Russian).
[72] KOTLYAR V. V., SERAPHIMOVICH P. G., SOIFER V. A. An iterative weight-based method for calculating kinoforms. Proc. SPIE Image Processing and Computer Optics, 1994, v. 2363, pp. 175–183.
[73] LU G., ZHANG ZH., YU F. T. S., TANONE A. Pendulum iterative algorithm

for phase retrieval from modulus data. Opt. Eng., 1994, v. 33, no. 2, pp. 548–555.

[74] FIENUP J. R. Phase-retrieval algorithm for a complicated optical system. Appl. Opt., 1993, v. 32, no. 10, pp. 1737–1746.

[75] KOTLYAR V. V., SERAPHIMOVICH P. G., SOIFER V. A. Weighting iterative method for calculating a kinoform. Opt. Spectrosc., 1995, v. 78, no. 1, pp. 148–151 (in Russian).

[76] VASILYEV E. D., KOTLYAR V. V., NIKOLSKI I. V., SOIFER V. A. Computer generated optical element for detecting the position. Proc. SPIE, 1993, v. 1983, pp. 1012–1013.

[77] KOTLYAR V. V., SERAPHIMOVICH P. G. Adaptive iterative method for kinoforms computation. Opt. Spectrosc., 1994, v. 77, no. 4, pp. 678–681 (in Russian).

[78] TIKHONOV A. N., ARSENIN V. Ya. *Methods for solving ill-posed problems*, Nauka Publishers, Moscow, 1979 (in Russian).

[79] BEREZNYI A. YE. Quasiperiodic optical elements. Comput. Opt., Moscow, 1989, no. 6, pp. 19–23.

[80] PAPOULIS A. *The theory of systems and transforms in optics*, Mir Publishers, Moscow, 1971.

[81] SIEGMAN A. E. Quasifast Hankel transform. Opt. Lett., 1977, v. 1, no. 1, pp. 13–15.

[82] SZAPIEL S. Rapid evaluation of the zero-order Hankel transform for optical diffraction problems. Opt. Appl., 1987, v. 17, no. 4, pp. 355–362.

[83] MAGNI V., GERULLO G., DESILVESTRY S. High-accuracy fast Hankel transform for optical beam propagation. J. Opt. Soc. Am. A, 1992, v. 9, no. 11, pp. 2031–2033.

[84] DOSKOLOVICH L. L., KHONINA S. N., KOTLYAR V. V., SOIFER V. A. Focusators into a ring. Opt. Quant. Electr., 1993, v. 25, pp. 801–814.

[85] KOTLYAR V. V., SOIFER V. A. Spatial filter for differentiating radially symmetric light fields. Lett. J. Techn. Phys., 1990, v. 16, no. 12, pp. 30–33 (in Russian).

[86] KOTLYAR V. V., SOIFER V. A., Rotor spatial filter for analysis and synthesis of coherent light field. Opt. Commun., 1992, v. 89, no. 2–4, pp. 159–163.

[87] BELANGER P., RIOUX M. Ring patterns of a lens–axicon doublet illuminated by a Gaussian beam. Appl. Opt., 1978, v. 17, no. 7, pp. 1080–1086.

[88] BRENDEN B. B., RUSSEL J. T. Optical playback apparatus focusing system for producing a prescribed energy distribution along an axial focal zone. Appl. Opt., 1984, v. 23, no. 19, p. 3250.

[89] MICHALTSOVA I. A., NALIVAIKO V. I., SOLDATENKOV I. S. Kinoform axicon. Optik, 1984, v. 67, no. 3, p. 267.

[90] KHONINA S. N., KOTLYAR V. V., SOIFER V. A. Focusator into longitudinal segment and multi-focal lenses. Comput. Opt., Moscow, 1993, v. 13, pp. 12–15.

[91] KHONINA S. N., KOTLYAR V. V., PHILIPPOV S. V., NIKOLSKI I. V., SOIFER V. A. Iterative methods for the kinoforms synthesis. OSA Proc. Int. Opt. Des. Conf., 1994, v. 22, pp. 251–256.

[92] FAINCHILD R. C., FIENUP J. R. Computer-oriented aspheric holographic optical elements. Opt. Eng., 1982, v. 21, no. 1, pp. 133–140.

[93] KOTLYAR V. V., PHILIPPOV S. V. Phase diffractive elements forming pregiven phase distributions. Opt. Las. Technol., 1995, v. 27, no. 4, pp. 229–234.

[94] VOLYNKINA Y. A., IBRAGIMOV E. A., USMANOV T. Determination of a class of laser beams featured by diffraction convergence. Opt. Spectrosc., 1992, v. 72, no. 6, pp. 1457–1483 (in Russian).

[95] KOTLYAR V. V., SOIFER V. A., PHILIPPOV S. V. Amplitude formers of wavefronts and amplitude lenses. Opt. Spectrosc., 1993, v. 75, no. 4, pp. 923–927 (in Russian).

[96] DURNIN J. Exact solutions for nondiffracting beam I. The scalar theory. J. Opt. Soc. Am., 1987, v. 4, no. 4, pp. 651–654.

[97] DURNIN J., MICELI J. J., EBERLY J. H. Comparison of Bessel and Gaussian beams. Opt. Lett., 1988, v. 13, no. 2, pp. 79–80.

[98] VALYAEV A. V., KRIVOSHLYKOV S. G. Modal properties of Bessel beams. J. Quantum Electron., Moscow, 1989, v. 16, no. 5, pp. 1047–1049.

[99] UNGER KH. G. *Planar and fiber optical waveguides*, Mir Publishers, Moscow, 1980 (in Russian).

[100] KOTLYAR V. V., SOIFER V. A. and KHONINA S. N. Phase optical elements for the generation of quasimodes of free space. J. Quantum Electron., Moscow, 1991, v. 18, no. 11, pp. 1391–1394 (in Russian).

[101] BELSKY A. M. Self-reproducing beams and their connection with nondiffracting beams. Opt. Spectrosc., 1992, v. 73, no. 5, pp. 947–951.

[102] PRUDNIKOV A. P., BRYCHKOV Y. A., MARICHEV O. I. *Integrals and series. Special functions*, Nauka Publishers, Moscow, 1983.

[103] KOTLYAR V. V., NIKOLSKI I. V., SOIFER V. A. Phase formers of Hermite beams. Opt. Spectrosc., 1993, v. 75, no. 4, pp. 918–922 (in Russian).

[104] GOLUB M. A., KARPEYEV S. V., KAZANSKY N. L., SOIFER V. A. Phase spatial filters matched to transverse modes. J. Quantum Electron., Moscow, 1988, v. 15, no. 3, pp. 617–618.

[105] GOLUB M. A., SISAKIAN I. N., SOIFER V. A., UVAROV G. V. Optical elements for analysing and forming a transverse-mode composition. Opt. Spectrosc., 1989, v. 16, no. 4, pp. 832–840 (in Russian).

[106] GOLUB M. A., SISAKIAN I. N., SOIFER V. A. Modans: new elements of computer optics. Comput. Opt., Moscow, 1990, no. 8, pp. 3–64.

[107] KOTLYAR V. V., NIKOLSKY I. V., SOIFER V. A. Phase formers of Hermite modes in diffraction orders. Lett. J. Techn. Phys., 1993, v. 19, no. 20, pp. 20–23 (in Russian).

[108] SMIRNOV V. I. *Handbook of higher mathematics*, v. IV, part I, Nauka Publishers, Moscow, 1974.

[109] VINOGRADOVA M. B., RUDENKO D. V., SUKHORUKOV A. P. *The theory of waves*, Nauka Publishers, Moscow, 1979.

[110] KORN G. A., KORN T. M. *Mathematical handbook for scientists and engineers: definitions, theorems and formulas for reference and review*, McGraw-Hill, New York, Toronto, London, 1961.

[111] DAMMAN J., KLOTZ E. Coherent optical generation and inspection of two-dimensional periodic structures. Opt. Acta, 1977, v. 24, pp. 505–575.

[112] MAIT J. N., BRENNER K.-H. Optical symbolic substitution: system design using phase-only holograms. Appl. Opt., 1988, v. 27, pp. 1692–1700.

[113] VELDKAMP W. B., LEGER J. R., SWANSON G. J. Coherent summation of laser beams using binary phase gratings. Opt. Lett., 1986, v. 11, pp. 303–305.

[114] SIMPSON M. J. Diffraction pattern sampling using a holographic optical element in an imaging configuration. Appl. Opt., 1987, v. 26, no. 9, pp. 1786–1791.

[115] Killat U., Rabe G., Rave W. Binary phase gratings for star couplers with high splitting ratio. Fiber Integ. Opt., 1982, v. 4, pp. 159–167.

[116] Arsenault H. H., Szoplik T., Macukow B. *Optical processing and computing*, Academic Press, San Diego, 1989.

[117] Seldowitz M. A., Allebach J. P., Sweeney D. W. Synthesis of digital holograms by direct binary search. Appl. Opt., 1987, v. 26, pp. 2788–2798.

[118] Turunen J., Vasara A., Westerholm J. Kinoform phase relief synthesis. Opt. Eng., 1989, v. 28, no. 11, pp. 1162–1167.

[119] Morrison R. L., Walker S. L., Cloonan T. J. Beam array generation and holographic interconnections in a free-space optical network. Appl. Opt., 1993, v. 32, pp. 2512–2518.

[120] Turunen J., Vasara A., Westerholm J. Stripe-geometry two-dimensional Dammann gratings. Opt. Commun., 1990, v. 74, pp. 245–252.

[121] Mait J. N. Design of binary-phase and multiphase Fourier gratings for array generation. J. Opt. Soc. Am. A, 1990, v. 7, no. 8, pp. 1514–1528.

[122] Doskolovich L. L., Soifer V. A., Kotlyar V. V. Phase diffraction gratings with preset intensity distribution between orders. Lett. J. Techn. Phys., 1991, v. 17, no. 21, pp. 54–57 (in Russian).

[123] Jahns J., Downs M. M., Prise M. E., Streibl N., Walker S. J. Dammann gratings for laser beam shaping. Opt. Eng., 1988, v. 28, pp. 1267–1275.

[124] Bobrov S., Kotletsov B., Turkevich Y. New diffractive optical elements. Proc. SPIE, 1992, v. 1751, pp. 154–165.

[125] Pommet A. D., Moharam M. G., Grann E. B. Limits of scalar diffraction theory for diffractive phase elements. J. Opt. Soc. Am. A, 1994, v. 11, no. 6, pp. 1827–1834.

[126] Soifer V., Golub M. Diffractive micro-optical elements with non-point response. Proc. SPIE, 1992, v. 1751, pp. 140–154.

[127] Doskolovich L. L., Soifer V. A., Alessandretti G., Perlo P., Repetto M. Analytical initial approximation for multiorder binary gratings design. Pure Appl. Opt., 1994, v. 3, pp. 921–930.

[128] Doskolovich L. L., Repetto M., Perlo P., Lavagno L., Sinesi S. Microfabricated multiorder diffraction gratings for fiber networks and broad to broad interconnects. Proc. ISHM 'The many aspects of microelectronics', 1994, Pavia, Italy.

[129] Karmanov V. G. *Mathematical programming*, Nauka Publishers, Moscow, 1986.

[130] Golub M. A., Doskolovich L. L., Kazanskiy N. L., Sisakyan T. N., Soifer V., Kharitonov S. A method of coordinated rectangles for calculating focusators into a plane domain. Comput. Opt., Moscow, 1990, no. 10, pp. 42–49.

[131] Kuhlov B., Ferstl M., Kobolla H., Pawlowski E. Holographic elements for optical interconnects at 1.5 μm. Proc. SPIE, 1992, v. 1751, pp. 66–75.

[132] Goncharsky A. V., Popov V. V., Stepanov V. V. *Introduction to computer optics*, Moscow State University Publishers, 1991 (in Russian).

[133] Doskolovich L. L., Kazanskiy N. L., Kharitonov S. I., Uspleniev G. V. Focusators for laser branding. Opt. Las. Eng., 1991, v. 15, no. 5, pp. 311–322.

[134] Soifer V. A., Doskolovich L. L., Golub M. A., Kazanskiy N. L. Diffraction investigation of focusators into a straight-line segment. Proc. SPIE, 1992, v. 1718, pp. 33–44.

[135] GOLUB M. A., DOSKOLOVICH L. L., SISAKYAN I. N., KHARITONOV S. I. Diffractive corrections in focusing laser light into a line-segment. Opt. Spectrosc., 1991, v. 71, no. 6, pp. 1069–1073 (in Russian).
[136] GOLUB M. A., DOSKOLOVICH L. L., KAZANSKIY N. L., KHARITONOV S. I., SISAKYAN I. N., SOIFER V. A. Focusators at letters diffraction design. Proc. SPIE, 1991, v. 1500, pp. 211–221.
[137] DALLAS J. M. Phase quantization – a compact derivation. Appl. Opt., 1971, v. 14, pp. 674–676.
[138] GOODMAN J. M., SILVESTRI A. M. Some effects of Fourier-domain phase quantization. IBM J. Res. Dev., 1969, v. 14, pp. 478–484.
[139] GOLUB M. A., DOSKOLOVICH L. L., KAZANSKIY N. L., SOIFER V. A., KHARITONOV S. I. Diffraction approach in multi-functional elements synthesis, Opt. Spectrosc., 1992, v. 73, no. 1, pp. 191–195 (in Russian).
[140] DOSKOLOVICH L. L., GOLUB M. A., KAZANSKIY N. L., SOIFER V. A., USPLENIEV G. V. Special diffractive lenses. Proc. SPIE, 1993, v. 1780, pp. 393–402.
[141] SOIFER V. A., DOSKOLOVICH L. L., GOLUB M. A., KAZANSKIY N. L., KHARITONOV S. I., PERLO P. Multifocal and combined diffractive elements. Proc. SPIE, 1993, v. 1992, pp. 226–234.
[142] SOIFER V. A., DOSKOLOVICH L. L., KAZANSKIY N. L. Multifocal diffractive elements. Opt. Eng., 1994, v. 33, no. 11, pp. 3610–3615.
[143] DOSKOLOVICH L. L., KAZANSKIY N. L., SOIFER V. A. Calculation of two-order focusators. Avtometrija, 1993, no. 1, pp. 58–62 (in Russian).
[144] GOLUB M. A., DOSKOLOVICH L. L., KAZANSKIY N. L., KHARITONOV S. I. Focusing of laser radiation onto line-round contours. Comput. Opt., Moscow, 1992, no. 12, pp. 3–8.
[145] GOLUB M. A., KAZANSKIY N. L., SYSAKYAN I. N., SOIFER V. A., KHARITONOV S. I. Diffractive calculation of the light field intensity in the vicinity of focal line. Opt. Spectrosc., 1989, v. 67, no. 6, pp. 1387–1389 (in Russian).
[146] GOLUB M. A., DOSKOLOVICH L. L., KAZANSKIY N. I., SISAKYAN I. N., SOIFER V. A., and KHARITONOV S. I. Calculation experiment with constant-intensity rectangular focusator of Gaussian beam. Comput. Opt., Moscow, 1990, no. 7, pp. 42–45.
[147] ALEKSOFF K. C., ELLIS K. K., NEAGLE B. D. Holographic conversion of a Gaussian beam to a near-field uniform beam. Opt. Eng., 1991, v. 30, no. 5, pp. 537–543.
[148] EISMANN M. T., TAI A. M., CEDERQUIST N. G. Iterative design of a holographic beam former. Appl. Opt., 1989, v. 28, no. 13, pp. 2641–2650.
[149] VORONTZOV M. A., SHMALGAUSEN V. I. *Principles of adaptive optics*, Nauka Publishers, Moscow, 1985.
[150] GOLUB M. A., DOSKOLOVICH L. L., KOTLYAR V. V., NIKOLSKI I. V., SOIFER V. A. Iterative-phase method for diffractively leveling the Gauss beam intensity. Comput. Opt., Moscow, 1993, no. 13, pp. 30–33.
[151] FEDOTOVSKY A., LEHOVEC H. Optimal filter design for annular imaging. Appl. Opt., 1974, v. 13, no. 12, pp. 2919–2923.
[152] PALCHIKOVA I. G., Phase-only synthetic hologram elements having increased focal depth. Comput. Opt., Moscow, 1989, no. 6, pp. 9–19.
[153] GOLUB M. A., DOSKOLOVICH L. L., KAZANSKIY N. L., SISAKYAN I. N.,

SOIFER V. A., KHARITONOV S. I. A method of conforming rectangles in calculating focusators into a plane domain. Comput. Opt., Moscow, 1990, no. 10, pp. 100–110.

[154] GOLUB M. A., KAZANSKIY N. L., SISAKYAN I. N., SOIFER V. A. Computational experiment with plane optical elements. *Optoelectronics, instrumentation and data processing*, 1988, v. 1, pp. 78–88, Allerton Press, New York, (translation of Russian journal 'Autometriya').

[155] KIPFER P., COLLISHON M., HAIDNER H., SHERIDAN J. T., SCHWIDER J., STREIBL N., LINDOLF J. Infrared optical components based on a micro relief structure. Opt. Eng., 1994, v. 33, no. 1, pp. 79–84.

[156] PETIT R. *Electromagnetic theory of gratings*, v. 22, Topics in Current Physics, Springer-Verlag, Berlin, 1980.

[157] MAYSTRE D. *Rigorous vector theories of diffraction gratings*, Progress in Optics, North-Holland, Amsterdam, 1984, v. 25, pp. 1–67.

[158] ANDERWARTHA J. R., FOX J. R., WILSON I. J. Resonance anomalies in the lamellar grating, Opt. Acta, 1979, v. 26, no. 1, pp. 69–89.

[159] MOHARAM M. G., GAYLORD T. K., Rigorous coupled-wave analysis of metallic surface-relief gratings, J. Opt. Soc. Am. A, 1986, v. 3, no. 11, pp. 1780–1787.

[160] KNOP K., Rigorous diffraction theory for transmission phase gratings with deep rectangular grooves, J. Opt. Soc. Am., 1978, v. 68, pp. 1206–1210.

[161] APOLLONOV V. V., BOCHKARYEV YE. L., ZASLAVSKY V. YA. Laser-beam splitter using diffraction grating. Quantum Electron., Moscow, 1979, v. 6, no. 3, pp. 615–618.

[162] HUGONIN J. P., PETIT R., CADILHAC M. Plane wave expansions used to describe the field diffracted by a grating. J. Opt. Soc. Am., 1981, v. 71, no. 5, pp. 593–597.

[163] BREKHOVSKIH L. M., Wave diffraction by a rough surface. J. Exp. Theor. Phys. (Moscow), 1952, v. 23, no. 3, pp. 275–304 (in Russian).

[164] VAN DEN BERG P. M., FOKKEMA J. T. The Rayleigh hypothesis in the theory of reflection by a grating. J. Opt. Soc. Am., 1979, v. 69, no. 1, pp. 27–31.

[165] SMOKY O. I., FABRIKOV V. A. *Methods of the theory of systems and transformations in optics*, Nauka, St. Petersburg, 1989 (in Russian).

[166] STREIBL N. Beam shaping with optical array generators. J. Mod. Opt., 1989, v. 36, pp. 1559–1573.

[167] BECKMANN P., SPIZZICHIO A. *The scattering of electromagnetic waves from rough surfaces*, Pergamon Press, Oxford, 1963.

Index